THE 8 PITFALLS OF ACCOUNTS PAYABLE AUTOMATION

Not Knowing Will Hurt You When Automating Accounts Payable

1st Edition 2011

Christopher Elmore

Copyright © 2011 by Christopher Elmore

All rights reserved. No part of this publication may be reproduced or transmitted in any form, by any means, electronic or mechanical, including photocopy, recording, or media information storage or retrieval systems now known or to be invented, without written permission from the writer, except by a reviewer who wishes to quote brief passages in connection with a review written for inclusion in an educational publication or radio or TV broadcast.

For ordering information, special discounts, bulk purchases or to book the author as a speaker, please visit www.newwayap.com

Edited by Rachel Stroud and Jan Skinner

Frist Edition 2011

ISBN-13: 978-1461039969

ISBN-10: 1461039967

To – Adella – Kyle – Emma – Ed – Ada – Mom and Dad

Table of Contents

Introduction 7

Chapter 1 Current Environment 15

Chapter 2 Vendor Selection 31

Bonus! 61

Chapter 3 Getting Organized 75

Chapter 4 The Project 97

Chapter 5 The Vendor 107

Bonus 2! 129

Chapter 6 The Training and Going Live 133

Chapter 7 Evaluation 147

Chapter 8 The Future 161

Conclusion 171

Acknowledgements 179

Glossary 183

Introduction

So... What's The Point?

Professional Paper Trackers

Traditionally an invoice is created by the vendor after the goods and services are delivered. The invoice is received by the person/department that is responsible for the expense. It is then sent to someone for coding against a budget and to categorize the expense for later reporting. Next the invoice is sent to yet another person who will enter it into the accounting system for cash management and payment. In a non-automated organization a higher number of invoices results in the need for more people to be added to the process, which ultimately lowers the company's bottom line. The headaches and challenges in AP comes when the coders and approvers in the above scenario are burdened by the volume and the paper being shuffled between multiple locations and approvers. If the people involved in the approval process are separated by geography, invoices will be lost and/or delayed, causing the process to breakdown. When this breakdown happens, companies start throwing people at the problem creating manual checks and balances to ensure the paper is in the right spot at the right time. This situation, over time, has created a class of professional paper trackers. Professional paper trackers are employees that, on a daily basis, conduct forensic investigations to find invoices that need to be paid. Paper is not traceable and, therefore, not reportable. This means a company never truly knows where invoices are at all times.

Technologies from the mid-1980s have helped by removing paper general ledgers; however the Accounts Payable world has lagged behind in automation and still mostly operates by passing paper. In my professional lifetime, I have personally witnessed a gradual shift in the adoption of automation in AP processes and the ultimate elimination of paper.

Before I Get To Far

A self-proclaimed computer nerd, I got my first computer in 1980; it was a PET and was driven by a cassette tape. Now this might not seem so wonderful, but as a kid I was mesmerized. I was awed by Space Invaders and I even got my name to scroll across the screen sideways. Then, I learned to hack the cassette tape by changing the code to make my ship in Space Invaders invincible to the bullets. (High Score... Very High Score!)

On a professional level, I have had the privilege of providing AP Automation to companies since 2000. I have worked with over 750 (and counting) AP departments, helping companies navigate their transition from paper to an automated AP platform. There were no guidelines, no best practices in this new frontier. I found myself inventing the rules as I went along. During my first implementation I had no project plan, no backup, and not much of a willing user base. I did, however, have an insider in the company that believed there was a need and, with technology, there was a way to eliminate data entry and paper. This was ground breaking, I thought, but simply not possible. However, the client was willing to pay for such an outcome so the CEO and President of the company I worked for said, "Yea – We can do that". This is where one might say, 'the rest is history'. After a lot of trial by fire, mad and happy clients, project plans, spreadsheets, PowerPoint's, xml and page refreshes, an industry has created a new automated process that is useful, stable, and affordable to every type of company.

However

I had an eye opening experience in New York City. I was working with a client who had automated one department and was interested in expanding the automation to the rest of the company. They hired a consulting firm to conduct its search on AP Automation companies. The sales pitch to the client from the consulting firm was: *We (the consulting firm) will determine your needs, and then we will assess the features of each AP Automation company's software. Once the needs are determined and the features assessed we will show which company has the product that best fits your (the client's) needs.*

That makes sense right? As this plan was outlined to me by the consulting firm's project manager, I said: "You are creating a problem by taking the client through this process because it's not all about features." She,

> *It is difficult to put "the ability to make a client successful" on a software selection checklist.*

with a little laugh, said, "No...it is all about the features". My attempt was to help her help the client understand that features alone would not make a successful AP Automation project. Realizing she was not open to my opinion, and believing that features alone would probably win me the deal, I dropped it. Much to my shock, I didn't get the expansion. Immediately after the decision, the calls started coming into my office from the client about wanting to know when their contract was up. A few months later the call came in wanting to know if they could switch to a month to month agreement after their initial agreement was up. Then, a few months later, the call came in requesting that we come back in to discuss our doing the original expansion. We discovered that the software implementation by the company they chosen crashed because the vendor of choice had no experience installing the necessary frame work to integrate to the client's accounting system. It is difficult to put "the ability to make a client successful" on a software selection checklist. Features are important because a company cannot afford to take step backwards in their process; however, features are only one of the many items that need to be considered when it comes to automating the accounts payable (AP) process.

This book is broken into the 8 pitfalls of AP Automation:

1. **Current Environment:** This chapter will help you understand what best practices are good for your company and how goals can be set appropriately. In order for a best practice to be useful, it has to be tailored to the company's current situation (especially as it pertains to technology).
2. **The Selection:** Identify the different types of vendors available for AP Automation and obtain a list of 10 questions you should ask your internal organization to determine the best vendor for your company.
3. **Getting Organized:** Learn how to organize your company to get ready for AP Automation. There are 8 questions to determine the best type of automation project for your company as well as best practices for setting up workflow.
4. **The Project:** Discover the 6 Steps to AP project "Greatness", how they are defined and what to expect when managing an AP Automation project.
5. **The Vendors:** Find out what to tell your vendors before you automate along with 15 questions to help you plan your vendor communications process.

6. **The Training:** We will explore how to effectively manage change management during the project and why end user training is not about the software; it's about dealing with change.
7. **The Evaluation:** Once you have completed your AP Automation project, this chapter will help you make sure that your goals for the project were achieved. Determine what lessons you learned and how to apply those lessons learned to future automation projects.
8. **The Future:** In this chapter, I offer my personal perspective on how AP Automation will evolve in the future as well as 5 technology ideas related to AP that you may want to consider as you evolve and grow into your "AP Automation Lifestyle".

As you continue to read, you will find this is not a technical book. There will not be any discussions of programming language or developer's techniques. This book was written for the person who is considering automating AP, but doesn't know where to start. This book was also written for the person (or company) who has automated their AP process and determined they need more or have a feeling that the new process is not going as well as it should.

I have written this book from the perspective of someone who has been in the AP trenches, having been personally responsible for managing 79 AP Automation projects (78 are still running successfully). This book is based on tried and true experience that I desire to pass on to you. Throughout my career I have carried out the duties of help desk phone support, project management, training (phone and in-person), consulting, sales engineering, sales management, product strategy, marketing, and even cleaning up around the office. Earlier I told you that I have worked with over 750 companies. Considering you are probably a numbers person, doing the math on 78 projects against 750 companies does not seem right; the difference in the numbers are those companies that did not become a client of mine because they either went with another company or the larger majority who decided to do nothing and continue to push paper. I believe those companies chose to do nothing due to good old-fashioned fear. It is true... changing a tried and true process can create fear. I believe fear in automating comes from not knowing what to do with displaced people that automation creates, and choosing the wrong technology provider. The biggest part of the fear comes from the newness of the technology and ideas. Generally people will not say, "I am scared to make a decision because I don't know anything about automated AP." They will say, "It cost too much." or "We don't have the personnel at this time to take on a project like this." or "Our current process is not so bad – we can live with it." Whatever the reason, the fear is real. Anyone in a modern business has heard the horror stories of a technology project that was poorly sold with huge promises of a life-changing event that didn't materialize. Some have heard it cost the company two arms and thousands of legs (there's an image), as someone put their reputation and job on the line then watched as it failed. That is why I have written this book. It is not as simple as picking a vendor, buying a product, and installing the product.

I have 4 goals in writing this book.

1. **Write a book people will be interested in reading.** Meaning the topic is timely and compelling.
2. **Write a book that people can read.** This is different than the first goal. The first goal relates to the topic the second goal relates to the style of writing or the voice of the book. I did not want to write a text book or a

white paper. I wanted to write a book in which my experience would speak to your experience in a meaningful way.
3. **Write a book that people would understand**. As with the second goal, this is not a text book or a white paper. It is also not a book that is written for developers. This book was written for the person that does not have programming skills and has no desire to learn software programming skills. The book was written with the intent to not talk about software at all, focusing instead on the process, change, and options.
4. **Write a book people can use**. Each chapter has questions for you to answer to better target a solution that fits your company's individual needs. One of the problems with emerging technology is the "one size – fits all" solution has not been established (over time technology can create more focused offerings), making it very important that recommendations are made in a customized way. With the combination of recommendations and information (knowledge) you will be better informed and more of an asset to your company in going from paper to automated AP.

Final Words

Relax, read and enjoy the ride, you will not regret the change. It can be done!

Chapter 1

Current Environment

Pitfall – Understanding Your Company's Current Environment Will Keep You Out Of The Pit.

Look For The Cat – It's Out Of The Bag

First things first... I am not an Accountant. Please, don't stop reading. In college I did take Accounting I and Accounting II, before changing my major to History. The fact that I do not have an accounting background has helped me tremendously. Early in my career, I had a client who was an early adopter and felt the software my company produced was lacking in many areas. He was trying to convince me to get the owners of my company to hire him, so he could "fix" the areas where the software was lacking. At the time I thought it was a good idea, after all he was a CPA. He listed out the things he would change, mostly around GL coding and business unit/cost center associations and fixed permission levels per role. I liked his ideas, so I went to my CEO. He said it was interesting but, at that time, he couldn't afford to pay me and a new person. Of course, I went back to the client and told him he got shot down by the CEO. Dishonest... a little, but here is the rest of the story; as time went by I began to realize that all of his suggestions were based on how his accounting department was operating and the "fixes" to our AP Automation tools were specific to his operation alone, not what was best for the industry. The client's extensive experience with one accounting system on the surface was a huge plus, but over time his ideas turned negative. To have success in the AP Automation industry, it is vital for a company to be the jack of all accounting systems, flexibility is the key. Another story, with a bit of a different twist, is about a new client who wanted functionality we didn't offer.

> Even though debits and credits are debits and credits in each business, the management of accounting practices vary

He was completely beside himself when he found out the software didn't do what he wanted it to. I can remember him saying, "This is basic accounts payable! I am shocked your other client's haven't rebelled, not having this." From my new client's perspective our perceived lack was something he had been doing for 20 years, completely clueless that it had zero impact within other companies. Here is my point: Even though debits and credits are debits and credits in each business, the management of accounting practices within every company is completely different. The bills have to be paid and profits have to be measured, but how the bills are paid and how the profits are measured is completely different. This is why having no previous accounts payable experience has been a benefit in my business. This is also why there is never a cookie cutter approach to automating accounts payable.

The Need To Define

Up to this point, I have assumed that you know what I am writing about when I refer to automated accounts payable. In the introduction chapter you read about the old school AP process. It is important that the assuming stops here, so we can proceed together talking about the same thing. The Random House Dictionary defines accounts payable as; "a liability to a creditor, carried on open account, usually for purchases of goods and services."[1] Meaning AP is the process of tracking and paying out money to suppliers or vendors. Automation is "the technique, method, or system of operating or controlling a process by highly automatic means, as by electronic devices, reducing human intervention to a minimum"[2] (Random House).

---

[1] http://dictionary.reference.com/browse/accounts+payable
[2] http://dictionary.reference.com/browse/automation

Automatic is the key word; "having the capability of starting, operating, moving, etc., independently." All together now, AP Automation is the independent moving of liabilities through approvals to payment. AP Automation eliminates needless steps and improves needed steps.

There are three main components to make AP Automation happen.

1. 100% Electronic Invoices
2. Event Driven Workflow
3. Reporting Layer to Track all Actions

100% Electronic Invoices: This statement is the most controversial. I used to play a little game when I did presentations by saying, early in the conversation, "The key to AP Automation is to get an electronic invoice from the vendor as soon as possible." I would not explain how do to that. After a few minutes, inevitably, a somewhat confused client would stop me and ask, "How do you do that?" To take away some of the mystery I would say, "The invoice can start as paper but needs to be scanned as soon as possible" (we will discuss this topic more in Chapter 5).

Event Driven Workflow: This type of workflow is the kind that does not need pushing or coaxing from a person. Once the invoice is associated to an approval process, subsequent invoices know their own steps to become fully approved. Early in my career I used to explain this as "smart paper" because the invoice, once electronic, knew the steps it needed to go through to be approved.

Reporting Layer to Track all Actions: "All actions" is the key statement. From top to bottom, entry to exit, all things big and small need to be tracked, for many reasons. One reason is to insure that everything the company needs to get done is done. Another reason is to optimize the processes. This is an important point because this is the first time you will know all actions of your process. In the paper driven world (technically) you will not know everything that is going on.

I had a client explain it as, "Once this is up and running I just want my guy to be sitting there pushing a few buttons now and then to keep invoice moving." I explained it would be more like 2001 A Space Odyssey when HAL (the computer) took over the entire ship, because the system will run on its own without someone pushing buttons to make it work. I was relieved when the client started laughing.

Enough With The Definitions

Now that things are all defined, the drama begins. In the early years of my career, I studied sales training. One thing I learned was the four stages of competence. The four stages of competence can help us understand where an AP Department is as it pertains to AP Automation.

Note: The word individual is used however it can be exchanged for company/department wherever necessary.

1. **Unconscious Incompetence** - The individual neither understands nor knows how to do something, nor recognizes the deficit, nor has a desire to address it.

2. **Conscious Incompetence** - Though the individual does not understand or know how to do something, he or she does recognize the deficit, without yet addressing it.

3. **Conscious Competence** - The individual understands or knows how to do something. However, demonstrating the skill or knowledge requires a great deal of consciousness or concentration.

4. **Unconscious Competence** - The individual has had so much practice with a skill that it becomes "second nature" and can be performed easily (often without concentrating too deeply). He or she may or may not be able teach it to others, depending upon how and when it was learned.

I have noticed most of the AP departments I have worked with around the idea of automation operate 80% of the time in with unconscious incompetence and the other 20% in conscious incompetence. The best advice I can give when a company is considering AP Automation, "Do not dismiss this new technology because it's new." If you go to dismiss a piece of technology like AP Automation, it needs to be dismissed based on the merits of your analysis. In this chapter I will help you move out of unconscious incompetence and into a decision that is right for your company in its current state. To do this, I am going to leverage the best practice framework that evaluates your company's current environment.

Best Practices for You! When I was in grade school I made the statement in class: "Practice makes perfect." One of the cool kids said, "No one is perfect so why practice". Thirty years later, while attempting to communicate with clients around the idea of "best practices", I found myself thinking about the cool kid's statement. An associate and I were attempting to develop a list of "best practices" that would be timeless and universal while communicating (with one sentence) how automated AP would operate in a modern business. We came to the conclusion it was not possible. A best practice compared to a practice is always in motion and is never constant.

There is no magical formula to success. The term, best practice, is a modern phrase that attempts to define the most effective and efficient way to complete a task or process. A best practice is a teaching tool that, over time, effort, and cost can be done in an optimized state. Business strategies such as Six Sigma, Lean Process, and TQM (Total Quality Management) have been conceived and sold on the principles of doing something the "right" way. However, there are two challenges with a best practice. The first challenge to a best practice is the environment. The environments of best practices are in a constant state of change. Change is a stick of dynamite to the ant hill which is a best practice (ah, to be a poet). Examples are hiring or firing, buyouts, sales, mergers, promotions, demotions, programs, rumors, plans, and trainings. More so than ever in corporate America, change is the only constant. The outcome of environmental change is that the principles the best practice are based on change, making the best practice no longer applicable. The second challenge with a best practice is innovation. Innovation is one of those positive negatives. It is good, but the practical applications of innovation are difficult to grasp, making it almost impossible to quantify within a company. Technology is the largest contributor to innovation but it is not the only one. Education, geography, social status and lifestyle are also major contributors. Every time there is a new innovation within a corporation, it changes the principles the best practice is based on. Here are a few examples where environment and innovation has affected a best practice.

Advances in medicine have given doctors the ability to extract a bullet from a victim's arm and within a few days they are engaged in physical therapy, whereas during the civil war the best practice was to amputate the entire arm immediately.

During the Victorian Era (1837 – 1901) it was thought that if a female was to go past a certain grade in school their brain would explode; whereas today there are more female scholars than ever before.

In 1954 Roger Banister broke the four minute mile despite reports from doctors that he would die if he tried to attempt it.

It was believed that a person would be unable to travel the speed of sound because, physically that person's molecules would not be able to handle the strain which is something Chuck Yeager proved wrong in 1947 when he became the first person to break the sound barrier.

In each of the examples above there was a perception, benchmark, or a mechanical limitation that was later broken by environment and/or innovation. Whether it was cultural, scientific or mechanical, after the innovation or environmental change the previous practice (best or not) was distorted and now seems so outdated it becomes laughable. Imagine going to the doctor with a broken arm and he says, "Yep it broken alright – looks like we'll have it cut it off at the shoulder." It's unfathomable, right?

Now, let's bring the idea of environmental change and innovation back to Accounts Payable. It wasn't very long ago that the personal computer was new to the work world. When I was in high school my Business Studies teacher spent an entire week on the "joys of punch cards" (if you do not know what punch cards are... Google it and have a good laugh). It's only been since the mid 1990's that accounting software has become available and affordable to every type of business. The innovation and environmental changes in the past 10 years alone have changed the best practices that came before them, as the next best practices will do 10 years from now.

One Step Forward - Two Steps to the Side

Identifying challenges associated with best practices will separate the popular notion of a best practice from a true best practice. Sales and Marketing machines overflow with ideas of best practices or best in class. *Best practice* has become a buzz phrase that is used in every pitch right beside scalable (see the bonus chapter) and sticky. Nowadays corporate leaders are creating mandates for their manager's to do more with less. Managers are being forced to find ways to operate more effectively and efficiently. Salespeople (who are also being pressed) are reaching more than ever to satisfy corporate leaders with assurance that their product or service is the "best". Although it may be true that the product or service helps, it does not automatically make it a best practice. My point: Best practices are based on independent needs and independent outcomes and no two companies are the same.

There are three steps in defining a best practice within your Accounts Payable department. The first step is AP health. How healthy is your Accounts Payable system.

8 PITFALLS OF ACCOUNTS PAYABLE AUTOMATION

First Edition 2011

Question – Section 1	Yes	No
1. Are invoices ever lost?		
2. Are invoices paid late?		
3. Is there a desire to capture discounts?		
4. Do statements go unrecognized?		
5. Do vendors call more than one person for an answer?		
6. Is the ratio between AP personnel to invoice more than 1:1000?		
7. Does processing cost more than $5 an invoice?		
8. Is there a "blame game" between AP and the rest of the company where leadership does not know who is wrong and who is right?		
9. Are invoices ever paid more than once?		
10. Is there more than one way to get an invoice to AP for entry?		

To calculate your total, start with 100 and subtract 10 for every question you answered was a **YES**.

Total from Section 1: _____

Question – Section 2	Yes	No
1. Are your invoices stored in document management software?		
2. Is payable data transmitted electronically to the account system?		
3. Is workflow software used for the approval process?		
4. Are invoices processed from a central location?		
5. Is information secure and visibility based on permissions?		

Next, add 5 points per question answered with a **YES** in Section 2. Total and add with the total from Section 1.

Total Score: _____

Standard grading scales apply:

        100-90 = A

        90-80 = B

        80-70 = C

        70-60 = D

        60 and Below = F

Now that you have obtained your health, let's find out your company's current state and direction. To find out, answer the following questions:

Question – Section 3	Yes	No
1. Is your company growing?		
2. Does your company want to grow?		
3. Has your company had a company software rollout in the last 18 months?		
4. Is your company currently (or should it be) cutting expenses?		

To determine your score for Section 3, add 2 points for each **YES**.

Total Score for Section 3: _____

(Note: for a printable version of the above chart/questions go to http://8pitfalls.com/ws/worksheet1.pdf)

The Best Practices - Custom Results:

The above questions will help craft what practices are best for your organization. This is to get an idea of the company's needs, direction, and current uses of technology. The results allow a best practice to become useful. Use the chart below to determine your company's category.

Description	Category
AP Health = A + Current State of 6 – 8	1
AP Health = A + Current State of 4 – 6	2
AP Health = A + Current State of 4 or below	3
AP Health = B + Current State of 6 – 8	2
AP Health = B + Current State of 4 – 6	3
AP Health = B + Current State of 4 or below	4
AP Health = C + Current State of 6 – 8	3
AP Health = C + Current State of 4 – 6	4
AP Health = C + Current State of 4 or below	5
AP Health = D + Current State of 6 – 8	4
AP Health = D + Current State of 4 – 6	5
AP Health = D + Current State of 4 or below	5
AP Health = F + Current State of 6 – 8	5
AP Health = F + Current State of 4 – 6	5
AP Health = F + Current State of 4 or below	5

Now What?

Congratulations! You have established a custom state and health for your AP department. Below is a list of best practices to consider implementing in your organization.

Category 5:

- Lower Cost
- Improve Controls
- Centralize vendor inquiries
- Eliminate re-work of invoices
- Find transitions in technology, compliance and/or audits and opportunities to improve change processes

Category 4:

- Create Visibility
- Common Documentation
- Single Data Entry
- Well Documented Business Rules
- Establish an exceptions handling process
- Create SLA or timers on each step of the process
- Eliminate paper filing
- Create an audit team for changes and updates to the approval system

Category 3:

- Electronic Invoices
- Single Audit Trail per invoice
- Develop an automated workflow process for approvals
- All invoices should come to a central location
- Automate the accruals process based on electronic information
- Send AP information to the accounting system in an importable electronic format

- Create an AP import to the accounting system that is capable of identifying errors and handling errors in the data
- Create a web-enabled system to approve and file invoices
- P-Cards
- Develop Electronic Signatures on Check
- Enable a role-based permission driven system
- Segregate approval data by entities
- Automate approval requests through a subscription email

Category 2:

- Automate checks for double entry
- Streamline your process so that only Approvers are in the workflow process - everyone else will be accommodated by reporting (see Chapter 3 for more details)
- Review the approval process and have the automated system make recommendations for improvement (This can be done through reporting on changes and updates)
- Create reverse statements to vendors
- Utilize ACH (Electronic Payment)
- Manage payment timing and terms for cash management
- Automate fraud prevention and detection
- Store invoices and process data in an offsite hosted data center

Category 1:

- Develop dashboards to track AP processes from multiple sources of data
- Write your own best practice chapter!

If you have not realized it yet, you need to crawl before you walk. With the information described above, the categories can suggest a good place to start. Keep in mind that the category suggestions are a starting point for your organization, and by no means am I implying that if you are in a low category you should not attempt AP Automation. Rest assured that, in my experience, an overwhelming majority of large to mid-size companies fall into categories 4 and 5 simply because the technology was unavailable at an appropriate price range. The rest of the book will concentrate on more general principles of AP Automation, so it is important to refer back to this chapter as you set your automation goals.

Chapter 2

Vendor Selection

Pitfall – When Selecting A Vendor Not Knowing Your ROI Will Put You In The Pit.

8 PITFALLS OF ACCOUNTS PAYABLE AUTOMATION

When I first got married, my wife wanted a dog. I looked in the paper and saw an ad for an "American Kerr Dog" $20. It was right in my price range. I was able to find the trailer park where the dogs were for sale, and met with the family selling them. How was I going to pick the right dog for my wife? As it turns out, when I got there all the puppies were looking at the man of the house wagging their tails. He said, "Watch this." and whistled to the dogs. Only one of the dogs turned her head and popped her ears up. I told him, "That's the one." Thirteen years later, after the dog passed away, I realized she was the best dog I've ever had. Attentive, easy to train, and fun ... she was the perfect dog.

Wouldn't it be great if selecting a service provider or partner was as easy as whistling? Having been on both sides of the selection process, I know it is an extremely difficult task. During the introduction, I told a story about a New York City client where the selection went wrong because they concentrated solely on the features of the product. In fact, many people think it is all about "features". That's exactly why and where many people go wrong. To figure out which service provider is the right one for you, you first have to do a little homework. There are three basic areas you need to evaluate and consider when making your selection: 1) cost per invoice 2) type of project based upon your desired ROI savings and 3) your organization's readiness for change. This chapter will walk you through the process of identifying and answering each of these areas for your company and help you stay out of the pit by not falling for the wrong one... solution provider, that is.

In a recent poll, the vast majority of people said that, "Technology is getting better and this year I want my company to invest in more technology to improve our Accounts Payable process". The problem is, people who made the previous statement do not know where to start, which vendor to use, whether to build a system themselves, the best way to achieve their technology goal or, if they automate, will their job be there in the morning? With technology and solution providers becoming more cost effective and competitive, AP is on the fast track to reinventing itself. This chapter will help you see if you are ready and able to change with the times. It will also provide you with a step-by-step process on how to make those changes.

There Is Hope After All - AP Can Be Fun!

Recently, a colleague of mine was working with a company in Minnesota. When asked for a reference on the fine work he had done, the client said that due to his (my colleague's) efforts and the outcomes of the project, AP has now become fun! **WHAT** – AP fun? The project was simply automating the approval process and creating open payables without manual entry into the accounting system. To capture the moment I created a diagram to illustrate the outcome (Fig. 1.1).

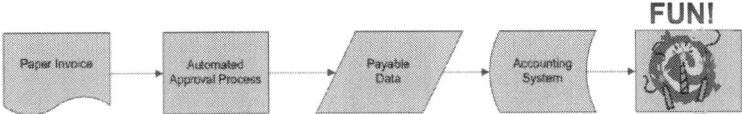

Figure 1.1

For those of you that operate in the "real" world of Accounts Payable...does the following story sound familiar?

8 PITFALLS OF ACCOUNTS PAYABLE AUTOMATION

First Edition 2011

An early start to the day, a little bit of traffic (for the west coast readers – everything is relative) a so-so parking space, holding a very heavy bag of stuff (paper) from the day before and then walking inside to a nice office environment and a hot cup of coffee. How am I doing so far? While getting your morning cup of coffee someone pokes their head into the break room door (not showing the rest of his body – as if he was hiding something) and says, "Good morning. Do you have that GL report for me? I need it today for my meeting." Inside you say, "Oh no – I forgot", but, due to years of training, you reply, "It's on my list of things to do this morning." He says, "Ok, when can I expect it?" (Inside) "Whenever it shows up." (Outside, from years of training) "45 minutes." You hurry back to your office to run the report, when you are stopped by the HR person. She asks you, "Did we pay Mark for his last day? I got a call from him and he said that he has not received his last check?" (Inside) "He didn't do anything while he was here so why should we pay him?" (Outside), "I'm not sure, I will have to check." You head back towards your office when the big boss makes eye contact with you (Uh-Oh); she says, handing you a piece of paper, "I need this invoice paid this afternoon, can you make that happen?" (Inside) "NOOOOO!" It must be for her brother.", (Outside) "Sure…I'd be happy to." You grab the paper with your little finger (the only part of you not carrying anything), and, of course, when you are two steps from your office your colleague who has only been at the company for two weeks says, "Good morning. The printer is not working and checks need to go out….any ideas?" You say, "Call IT." Your colleague says, "I did and

> *Good morning, the printer is not working and checks need to go out….any ideas?*

8 PITFALLS OF ACCOUNTS PAYABLE AUTOMATION

they told me to ask you to fix it". (Inside) "I'll fix it alright." (Outside) "Ok, hold on a minute; let me put my stuff down."

Now, you are finally in your office as the phone is ringing. You see the number scrolling across your phone and recognize they've called you several times and never left a message... curiosity gets the best of you so you pick it up and say, "Accounts Payable, how can I help you?" The voice on the other line says, "I am calling about a past due invoice in the amount of $7.52 that is now 32 days past due. Can you tell me when you're going to pay the invoice?" (Inside) "I just saw that invoice in the trash where it belongs." (Outside) "What's the invoice number? I will look it up and call you back." After taking the vendor's information down you look at the stack of invoices that need to be filed and entered into the accounting system... you do this six or seven times – back and forth like a crazy person. Then, paralyzed with indecision, you look at the clock and its 8:35 (Inside) "3 hours and 25 minutes until lunch – I wonder what the drink specials are?" (Outside) "3 hours and 25 minutes until lunch.... I need a drink!"

So, let's recap: Officially your AP Hero's day starts at 8:30; at least that's what the paycheck is based on. This AP person is not more than 5 minutes into their day and their immediate list of things to do reads like this (in no particular order because they all have to be done at the same time):

1. Run and send GL Report
2. Print GL Report
3. Get GL Report to person for meeting
4. Find out if Check has been cut and sent to an old employee
5. Email response back to HR on status of the check
6. Check invoice from CEO for approval and validity

7. Enter invoice for CEO
8. Do special check run for CEO
9. Report back to CEO that the invoice was entered and the check was sent
10. Fix the printer
11. Find out from your boss if IT should have fixed the printer.
12. Talk to IT about their service to the accounting department (in a non-threatening way)
13. Try to get over the paranoia of confronting IT
14. Research the $7.52 invoice
15. Call the vendor back on the status of the $7.52 invoice
16. Explain to the vendor that you are not the right person to call the next time they have a question about payment
17. Enter the daily stack of invoices into the AP system
18. Match and file invoices that were cut yesterday

That's 18 items needing to be done as soon as possible. Out of those 18 items, only 2 items were predictable. Before the day started, you knew you were going to enter the daily stack of invoices and file the invoices that were entered the previous day. For argument's sake, the first three items could/should have been predicted; however it is also arguable that the request for the report never got to you in a meaningful way so you could have planned for it. Here is the point: the typical AP person has no idea what tasks are going to pop up to derail them from what they had originally planned to do; therefore causing a typical AP person's life to be very <u>reactive</u> instead of <u>proactive</u>.

What Is Holding You Back From Change?

When you look back on the last 20 years, there have been tremendous changes in how technology is used in business, most notably, based on the rise of the personal computer and the internet. In the last 10 years, accounting systems have grown and developed into very powerful systems. And even more recently with the widespread addition of the internet, companies and the technologies that drive those companies have provided many opportunities with "real" tangible value for accounts payable process improvement. Here is a short list of the technologies available today for the AP Department:

- Document Scanning
- Document Storage
- Electronic Data and Data Exchange
- Automatic Capture of Invoice Data
- Workflow Process
- Compliance and Audit Capabilities
- Reporting Across Multiple Applications

To portray accounts payable as a forgotten department would be a gross misrepresentation. However, an argument could be made that the AP department is viewed as a pure expense of the company and not entitled to be improved through an investment in technology. I have had many discussions with colleagues, consultants and clients on why this is the case. My personal view is that organizations have an extremely difficult time understanding the impact/outcome of an automated accounts payable department. Therefore there is a lot of "lip service" given at conventions and trade shows on companies automating, but when the attendees go back to their regular life they slip right back into the list of 18 things that has to be done ASAP. The reason for this is FEAR. I had a mentor that once told me fear stands for **F**alse **E**vidence **A**ppearing **R**eal. I believe that once the AP staff person returns to the office, they start to fear that automating will cause their lives to be worse or the technology will be more complicated than they can handle. Their list of 18 will become 57 or, even worse, they will be "reassigned" within the company. Once the fear is overcome, the reality of AP Automation is so much better than they could have ever imagined.

Overcoming the Fear

Here is the true reality of technology.... it does nothing. Well that's not completely true. A better way of stating it is technology does nothing without people. The myth about technology is that it is some type of strange human-like-ever-evolving-beast that will eventually take over the world. In fact, some people think computers will get to the point where they can do everyone's job. It is true that advances in technology will make some jobs or aspects of jobs obsolete. People who do not or cannot change put themselves in harm's way as technology is introduced, while others are propelled higher in their organization due to their embrace of and desire to learn more about the new technology.

In an attempt to dispel the fear of change, there are three questions you should ask yourself and/or others in your organization:

What would you do with your time if you did not have to...

1. Open and sort the incoming invoice?
2. Enter payables into the accounting system?
3. File invoices?

The challenge with these questions is there are many answers depending on the person asked. The CEO may say, "I would close the entire department..." Then who would send out the checks? The remote people might say... "I didn't know you had to enter payables data into the accounting system." In my opinion, this question is best answered by the leadership of the Accounting or Accounts Payable department.

Below are a few examples of what Accounting or the AP department can do with their additional time.

- Concentration of Cash Management
- Improve Vendor Relationships
- Implementation of Vendor Discounts
- Purchase Card Program
- Adding and Managing Contracts with Vendors

But, before you can add value to your company as in those ideas listed above, you need to answer the three questions fully. By answering those questions, collectively, you will be able to determine a single answer to your software dilemma that defines what technology you choose, which type of vendor you will use, and, most importantly, measure how great your return on investment (ROI) will be. Below is a six step process to drive to your single answer:

1) Outline your accounts payable process.
2) Assign a dollar value to each step in the process.
3) Calculate your cost per invoice (CPI).
4) Determine the type of technology vendor based upon your desired savings.
5) Eliminate each step that will be covered by automation (ex. receiving, sorting, routing the paper, data entry, filing, etc.).
6) Calculate your ROI.

I know this exercise is easier written than actually done. For that reason, I have provided an outline to help you step by step through the process. Please don't tell your consulting company that I have made this an easy, straight forward process – this goes for big bucks in the consulting world.

Determine your Cost per Invoice (CPI)

To begin the process, I thought it might be helpful to see what a completed questionnaire would look like, and how the results are calculated. The first section below has the questions in "bold black" and the example answers in "lite black". At the end of this example section is a blank questionnaire for you to enter your information.

Example Section:

1) **When invoices are sent from the vendor where do they arrive?** Each invoice goes to the location where the original expense was created. Most locations are outside of our corporate office. There is a mailroom in each location that sorts the mail and delivers it to the various mail slots for each department. Invoices represent about 60% of overall mail traffic and we have one dedicated person per location or regional office for the mailroom. Time – 6 hours per day or 24 hours per week per person per regional office or 96 hours per month

2) **When they arrive, who (role) opens the mail and what do they do once it is opened (list each possible path of the invoice)?** The assistant to that department opens the mail and puts each invoice in its proper mail slot for the correct

manager. Time – 5 hours per week /20 hours per month over 15 departments

3) In each possible path listed above, what does the person do? (Note: This includes research, trouble shooting and resolving budget conflicts.) The manager checks the budget and codes the invoice. After they confirm the work has been done or the goods have been delivered, they sign off on the invoice. Time – 14 hours per week / 56 hours per month x 15 managers

4) When that person is done, where is the invoice sent and how is it sent? The manager gives it to the assistant to either mail to corporate or the regional manager's office for additional approvals. Time – 6 hours per week /24 hours per month x 15 departments

5) When the invoice arrives at the next location what happens, who touches it, and what do they do? Invoices that are sent to corporate are divided by the receptionist of each business unit and given the to the appropriate AP person for entry into the accounting system. Time – 8 hours per week per receptionist /32 hours per month

Note: Depending on your organization, you may have to repeat questions 3-5 several times. It is important that you capture a complete list of all the steps in the process that touch an invoice in order for your ROI to be accurate. When the invoices are sent to the regional manager, the regional's assistant opens the interoffice mail and puts the invoices on the regional manager's desk for approval. Once the invoices are approved, the regional manager gives them back to their assistant to be sent to corporate for entry into the accounting system. Time – Assistants 5 hours per week /20 hours month and the Managers - 10 hours per week /40 hours per month across 8 regional offices.

6) Once the invoice is fully coded and approved, how many invoices are entered into the accounting system and how much time does it take to enter the invoices? (Note: If you have difficulty obtaining the average number of monthly invoices, you can base it on 1.3 x number of checks written each month.) 11000 invoices are processed per month. Each AP person enters a daily average of about 110 invoices into the system. There are 15 clerks, but only 5 of them enter data full-time. Time - 40 hours a week /160 hours a month for 5 clerks

7) Are invoices matched to the checks as backup for check signing? If yes, how much time is spent and who does that? Yes, this is full-time job for 3 clerks. Time – 40 hours per week /160 hours per month for 3 people

8) Who files the invoices and how much time is spent filing? Each AP clerk is responsible for maintaining a temporary filing system for each invoice they enter and that time is included in answer 6 above. Once the check run is complete, the invoices are retrieved by another group of AP clerks and filed. Part-time job for 3 clerks. Time – 20 hours per clerk per week / 80 hours per month for 2 people

9) Before the invoice is filed are copies made? If so, by whom and how many times? Yes, copies

of invoices are made for each location and sent to each location weekly. Other half of filing clerks jobs. Time – 20 hours a week / 80 hours per month for 3 clerks

10) Who answers the call from vendors asking for status on their payments? How many hours a week do they spend tracking down the invoice and getting back to the vendor? 3 clerks answer these calls full-time. Time – 40 hours a week per clerk/160 hours per month. (Note: Make sure that once you have accounted for all your AP Clerks' time that you compare the total time to an overall 40 hour week so that you don't over/understate the hours.)

11) What does your company spend in sending invoices back and forth between locations on a monthly basis? $2,950

12) What does your company spend in offsite storage costs for invoices per month? $4,250

13) What does your company spend in late fee penalties per month? $20,500

Hourly Rates per Role:

In the section provided below, list each role and their associated hourly wage as outlined in the questions above. (Note: Divide the annual salary by 2080 to obtain the hourly rate of the role.)

Role: Mailroom per hour	Wage $13.10

Role: Assistant Manager per hour	Wage $16.82
Role: Manager per hour	Wage $29.80
Role: Assistant per hour	Wage $16.85
Role: Regional Assistant per hour	Wage $15.80
Role: Receptionist per hour	Wage $13.20
Role: Regional Manager per hour	Wage $52.88
Role: AP Clerk per hour	Wage $17.80

Math:

Now, pull it all together and do the math for each section. Multiply the numbers of hours per month by the number of staff and the wage per hour for the staff member to get a cost per line per month.

Then total up your costs and divide that total by the number of invoices that are managed each month to determine your Cost per Invoice (CPI):

Step	Hours/Month	Role	# Staff	Wage/ Hour	Cost per Step
1	24 hours per week or 96 hours per month per region	Mail Room	8	13.10	$10,060
2	5 hours per week or 20 hours per month per department	Assistant Manager	15	$16.82	$5,046
3	14 hours per week or 56 hours per month for 15 departments	Manager	15	$29.80	$25,032
4	6 hours per week or 24 hours per month for 15 departments	Assistant	15	$16.85	$6,066
5	8 hours per week or 32 hours per month	Receptionist	8	$13.20	$3,379
	Assistants - 5 hours per week or 20 hours per month for 8 regions	Regional Assistants	8	$15.80	$2,528
	Regional Managers - 10 hours per week or 40 hours per month for 8 regions	Regional Managers	8	$52.88	$16,921
6	40 hours a week or 160 hours a month per clerk	AP Clerks	5	$17.80	$14,240
7	40 per week or 160 hours per month per clerk	AP Clerks	3	$17.80	$8,544
8	20 hours per week or 80 hours per month per clerk	AP Clerks	3	$17.80	$4,272
9	20 hours per week or 80 hours per month per clerk	AP Clerks	3	$17.80	$4,272
10	40 hours week or 160 hours per month per clerk	AP Clerks	4	$17.80	$11,392
	Subtotal Labor				$111,752
	Benefit Burden Rate (Ave 23%)				$25,703

	Total Labor Costs	$137,455
11	Monthly Fed Ex or Mail Cost	$2,950
12	Monthly Off Site Storage Cost	$4,250
13	Monthly Late Fees	$20,500
	Total Cost per Invoice (CPI) per Month	$165,155
	Total Invoices per Month	11,000
	Cost per Invoice (CPI)	$15.01

(Note: for a printable version of the above chart/questions go to http://8pitfalls.com/ws/worksheet9.pdf)

Your Turn

Step 1: Outline your accounts payable process. In this step you will cover your current process, exceptions and other special situations. Make sure to include the amount of time needed to complete each step. Be sure to capture each step in the process to ensure that you get an accurate cost per invoice.

Questions:

1) When invoices are sent from the vendor where do they arrive?
2) When they arrive, who (role) opens the mail and what do they do once it is opened (list each possible path of the invoice)?
3) In each possible path listed above, what does the person do? (Note: This includes research, trouble shooting and resolving budget conflicts.)
4) When that person is done, where is the invoice sent and how is it sent?

5) When the invoice arrives at the next location what happens, who touches it and what do they do?

Note: Depending on your organization, you may have to repeat questions 3-5 several times. It is important that you capture a complete list of all the steps in the process that touch an invoice in order for your ROI to be accurate.

6) Once the invoice is fully coded and approved, how many invoices are entered into the accounting system and how much time does it take to enter the invoices? (Note: If you have difficulty obtaining the average number of monthly invoices, you can base it on 1.3 x number of checks written each month.)

7) Are invoices matched to the checks for backup for check signing? If yes, how much time is spent and who does that?

8) Who files the invoices and how much time is spent filing?

9) Before the invoice is filed are copies made? If so, by whom and how many times?

10) Who answers the call from vendors asking for status on their payments? How many hours a week do they spend tracking down invoices and getting back to the vendor?

11) What does your company spend in sending invoices back and forth between locations on a monthly basis?

12) What does your company spend in offsite storage costs for invoices per month?

13) What does your company spend in late fee penalties per month?

14)

Math:

Using the grid below:

1) From the information that you've collected during your process above, please enter the hours/month, the role and the number of staff at each step.
2) Determine the Wage/Hour and enter below based on each role listed below.
3) Calculate the cost per step in your process and total your costs to determine the Total Cost per Invoice per Month.
4) Divide that total by the number of invoices each month to determine your Cost per Invoice (CPI).

Step	Hours/Month	Role	# Staff	Wage/Hour	Cost per Step
1					
2					
3					
4					
5					
6					

8 PITFALLS OF ACCOUNTS PAYABLE AUTOMATION

First Edition 2011

7					
8					
9					
10					
	Subtotal Labor				
	Benefit Burden Rate (Ave 23%)				
	Total Labor Costs				
11	Monthly Fed Ex or Mail Cost				
12	Monthly Off Site Storage Cost				
13	Monthly Late Fees				
	Total Cost per Invoice (CPI) per Month				
	Total Invoices per Month				
	Cost per Invoice (CPI)				

(Note: for a printable version of the above chart/questions go to http://8pitfalls.com/ws/worksheet2.pdf)

Determine your Type of Project based upon your desired ROI Savings

Congratulations! You've done it! Now that you have captured your CPI, it is time to determine which type of product best suites your company based on the ROI that will return to your company. There are 5 major categories of solutions to automate AP and because each technology vendor does not supply the same type of labor savings on the various areas of your business, each of them offers a different level of ROI. This is an extremely important step in your ROI analysis. Here are the five types of technology vendors:

1. Self-Scan – Self Storage Document Management
2. Self-Scan – Vendor Storage Document Management
3. Self-Scan – Self Storage AP Automation
4. Outsource Scan – Self Storage AP Automation
5. Outsource Scan – Vendor Storage AP Automation

Confused yet? Hopefully the distinction between Document Management vs. Automation has taken you by surprise. And hopefully your about to justify reading all of this as you grasp the understanding of this very fine, but confusing point. Document management is simply electronically storing a file (such as an invoice). Automation is the creation of an electronic process (that usually replaces a manual process). There is a huge difference between these two ideas and an even larger variance in the ROI.

AP Automation vs. Document Management

Before diving into these subjects, there two things you need understand. (1) AP Automation and electronic storage are two very different things that get confused on a regular basis during the selection process. As outlined in the introduction, AP Automation eliminates steps and improves steps in AP using technology that contains three main components: electronic invoices, a workflow process and electronic storage. Electronic storage on the other hand is defined as the conversion of a paper document (ex. invoice) to an electronic file that can be stored and later retrieved.

The primary difference between AP Automation and electronic storage comes down to amount of time and labor savings a company will experience after implementing those systems.

> The primary difference between AP Automation and electronic storage comes down to amount of time and labor savings

With electronic storage, your company will still have to manage the paper requiring someone to open the mail, scan the document, shred the document, etc. Alternatively, using AP Automation the paper goes away, completely. With AP Automation you are actually eliminating tasks within your team (such as entering the AP data into your accounting system). Document Management systems do not eliminate tasks; they simply turn paper into electronic storage. Both systems allow you to store and later retrieve the documents. So, as it relates to the ROI and using a Document Management system, this type of system forces you to maintain the same cost structure within your AP department and adds the cost of the storage system on top. Also be aware that electronic storage vendors attempt (on a regular basis) to offer solutions for AP Automation and, if you head down that road, realize you're not really getting rid of the paper; you are just automating the process of storing documents and may not be saving as much as you initially intended.

Picking a Software Platform

As a software platform, Software as a Service (SaaS) has become a more dominant force for consideration over the classic Application Service Provider (ASP) and self-hosted client server models. To help you better understand the differences between these three software platforms; let's start with a brief explanation of each model. With a self-hosted or client server model, a company typically uses internal resources to install the software (one instance per each user base) and provide ongoing software maintenance as well as manage the internal systems, including storage. ASP is similar to the hosted model, with the only difference being that the ongoing maintenance and storage is handled by a third party hosting company. SaaS is similar to ASP in that the data and systems are maintained and stored by a third party; however different from the ASP model, SaaS uses a single instance of the software to manage multiple user bases. Keep in mind that there is a difference in the types of software used for AP Automation software and storage as well as software platforms (ex. Hosted, ASP and SaaS).

Self-Scanning vs. Outsourced Scanning

Other factors that plays into ROI are the labor and hardware savings and differences between self-scan vs.

outsource scan. Self-Scanning options provide you the ability to do your own scanning and storage. So, by using those solutions, your current AP department staffing would remain the same and you would need to invest in the appropriate level of scanning equipment for your volume as well as assign staffing (or add staff if your current people are swamped). By outsourcing the scanning, all the staffing related to touching paper goes away completely (i.e. Opening the mail, scanning, etc.) There is no need to purchase or maintain any expensive scanning equipment.

To give you an idea of the differences between the various technology options, we have listed below the percentage of potential savings by vendor:

Technology Vendor	Potential ROI Savings
1. Self-Scan – Self Storage Document Management	10%
2. Self-Scan – Vendor Storage Document Management	15%
3. Self-Scan – Self Storage AP Automation	25%
4. Outsource Scan – Self Storage AP Automation	30%
5. Outsource Scan – Vendor Storage AP Automation	55%

Using the original example ROI listed above; here is a different way of illustrating the impact of the various technology options:

Total Cost per Invoice (CPI) per Month	$165,155
Total Invoices per Month	11,000
Cost per Invoice (CPI)	$15.01

Applying Savings Percentages to the ROI Example:

Technology Vendor	Savings	Savings	New CPI
1. Self-Scan – Self Storage Document Management	10%	$16,515	$13.51
2. Self-Scan – Vendor Storage Document Management	15%	$24,773	$12.76
3. Self-Scan – Self Storage AP Automation	25%	$41,289	$11.29
4. Outsource Scan – Self Storage AP Automation	30%	$49,546	$10.51
5. Outsource Scan – Vendor Storage AP Automation	55%	$90,835	$6.75

The trick to this entire evaluation is to find a software/service provider that will fit within one of these savings models and cost less than your estimated ROI. Here is an example: The numbers above are conservative estimates. I have personally seen a cost per invoice (CPI) as high as $45.00. Most of the projects I have worked on have an average CPI between $15.00 and $22.00. The decision to automate or not becomes easier and easier as the CPI goes up. Let's say that a company is interested in Option 1 – Self Scan – Self Storage Document Management. They would need to find a software/service provider that would be able to handle a volume of 11,000 invoices per month for less than $16,515. Be careful to compare apples to apples when it comes to pricing.

Software companies may offer their annual maintenance at that price, but you also have to factor in the set up and implementation cost. True, sometimes the set up and implementation costs can be justified as an investment; however, it doesn't make sense to spread out those costs for more than a year. Please note that as a rule of thumb the set up and implementation costs are reversed in scale to the above savings module. Meaning, Option 5 – Outsource Scan – Vendor Storage AP Automation will normally have the highest setup and implementation and where Option 1 Self Scan – Self Storage Document Management will have the lowest.

Determine your Organization's Readiness for Change – the "X" Factor

The last factor to consider when evaluating electronic AP is your organization's ability to accept and manage change. The effect of change on an organization is dramatic. Throughout my experience with AP Automation, change management has always been the "X" factor when it came to the success or failure of a project. Change is such an emotionally charged issue that it is the one aspect of the project that has the ability to keep its sponsors up at night. Change has kept me up too, to the point that I dedicated an entire chapter to the subject (Chapter 6 – The Training and Going Live). To help you correlate the impact that change management will have on the success of your AP Automation process, I have devised a quick and easy Change Factor Scale, so you will be able to see if electronic AP is for you.

Change Factor Questions: (Answer the following questions)	Yes/No
1. Has your organization gone through an accounting system conversion (going from one system to a new system) within the last 3 years?	
2. Was the conversion on time?	
3. Was the conversion under budget?	
4. Did employees retain their employment with the company due to that conversion (no one fired or quit)?	
5. Has your organization performed an accounting system upgrade (new version or added modules) within the last year?	
6. Was the upgrade on time?	
7. Was the upgrade on budget?	
8. Did employees retain their employment with the company due to that upgrade (no one fired or quit)?	
9. Is your accounting system on spreadsheet or on paper?	
10. Is your accounting system hosted by the software provider?	

Change Factor Calculations:

For questions 1-8, for every Yes add a factor of 1

Total: _____

For question 9, for a Yes subtract a factor of 5

Total: _____

For question 10, for a Yes add a factor of 5

Total: _____

Grand Total: _____

Change Factor Scale:

Factor	Suggested Outcome
9-13	Highly suggest full automation
7-8	Suggest full automation
4-6	Proceed with caution
2-4	Change at this point may do more harm than good
0-1	Not a good idea
Below 0	Do not do anything

(Note: for a printable version of the above chart/questions go to http://8pitfalls.com/ws/worksheet3.pdf)

Bring It On Around:

So here is the dramatic conclusion. There are three things to consider when selecting the right AP Automation vendor:

(1) Cost per Invoice (CPI) and your desired Return on Investment (ROI).
(2) Finding the Type of Software/Service that delivers ROI your company demands.
(3) Change management and how it impacts your company.

Once each of these items have evaluated and measured, the decision has been made for you. The next step is to convince the purse string people to pull the trigger. If you are the purse string person... go ahead, and use this information as justification to your team, department, or board. If you are not the purse string person use this information as your pitch to improve accounts payable and your organization.

One Last Thought:

Remember at the beginning of this chapter we met an accounts payable manager. We last left our hero counting the hours until lunch. Fast forward a year into the future with her and look at her revised 18 items list in a fully AP automated world:

1. ~~Run and send GL Report~~
2. ~~Print GL Report~~
3. ~~Get GL Report to person for meeting~~ – Create an automated report sent via email to the right person at the right time.
4. ~~Find out if Check has been cut and sent to an old employee~~ – Enable the requester to run a search and find the information without any help from AP.
5. ~~Email response back to HR on status of the check~~ - Enable the requester to run a search and find the information without any help from AP.
6. ~~Check invoice from CEO for approval and validity~~ – Workflow approval process will automate this.

7. ~~Enter invoice for CEO~~ - Check request entry by the CEO.
8. ~~Do special check run for CEO~~ – Check request entry by the CEO.
9. ~~Report back to CEO that the invoice was entered and the check was sent~~. – CEO could look up online to check the status of the payment.
10. Fix the printer – Try ACH payment.
11. Find out from your boss if IT should have fixed the printer. – Good luck!
12. Talk to IT about their service to the accounting department (in a non-threatening way) – Can't help you here!
13. Try to get over the paranoia of confronting IT – Sorry you're on your own!
14. ~~Research a $7.52 invoice~~ - A vendor login will allow the vendor to find their own answer.
15. ~~Call the vendor back on the status of the $7.52 invoice~~ – A vendor login will allow the vendor to find their own answer.
16. ~~Explain to the vendor that you are not the right person to call the next time they have a question~~. – Refer the vendor to a web log in through a recorded phone message, website posting, letter and/or statement in the remittance envelope.
17. ~~Enter daily stack of invoice into the AP system~~ – Outsourced invoice capture and electronic invoices replace the need to enter invoices into AP.
18. ~~Match and file invoices that were cut yesterday~~ – Invoices are electronically stored.

New List:

(1) Concentrate on your core responsibilities.
(2) Continue to leverage technology to do the things that can be automated.
(3) Enjoy your job!

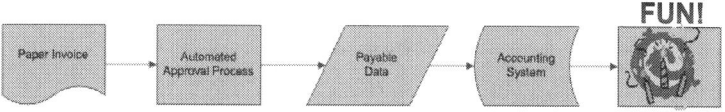

Making your job more enjoyable may not be the primary goal of you and your organization, however becoming more productive by leveraging technology certainly can make you a hero to the leadership of your company. On the other hand, selecting the wrong service provider for the job or creating unrealistic goals with that service provider can make you a zero in your organization. By better understanding your internal organization, the different types of scanning, storing and process options and the impact these can have on your company, you will be more confident in selecting the best type of service/solutions provider for your company and ensure you do not fall into the pit by choosing the wrong partner.

Bonus! Scalability

Who's Scale Measures Scalability?

Since 2000 I have had the great pleasure of sitting in as well as presenting several hundred (closer to a thousand) software demos. For those of you who do not know this, a software sales engineer or sales person is required by the unwritten law of software to use the word scalability in every meeting, pitch, demo or phone call. If they don't they will be black balled from the world of software sales. When I was demonstrating software I would get to the scalability point and say, "Ok, here it comes – the one word that every consultant has to say." I would pause to make sure the air was thick with anticipation. Then I would say... "It's scalable!" There would be groans and a little bit of a letdown. Then I would explain why it is scalable. However, in this world of software and software sales, I think sales representatives don't know what it really means and why it is important. Even if the sales person knows what scalability is, they take it for granted that the prospective client knows the term or knows the true impact of the term.

In this section you will find out the what, why and the how of scalability and what you need to know before you weigh in on its impact before you automate you accounts payable process.

Here is how the dictionary defines scalability:

"The ability of something, especially a computer system, to adapt to increased demands[3]."

That's a great working definition; the dictionary went on to explain:

---

[3] http://dictionary.reference.com/browse/scalability

For example, a central server of some kind with ten clients may perform adequately but with a thousand clients it might fail to meet response time requirements. In this case, the average response time probably scales linearly with the number of clients, we say it has a complexity of O(N) ("order N") but there are problems with other complexities. E.g. if we want N nodes in a network to be able to communicate with each other, we could connect each one to a central exchange, requiring O(N) wires or we could provide a direct connection between each pair, requiring O(N^2) wires (the exact number or formula is not usually so important as the highest power of N involved).[4]

My head hurt just a little from writing that definition. I imagine yours may be from reading it. However, the definition explains that the more data (of strain) you put on a system without the appropriate capacity the system will not work to capacity. (Furthermore) The problem with scalability as it pertains to an automated AP environment is the above definition is only one-third of the overall scope, which is "network" scalability. I have created a way to analyze each of the three aspects, which we will call "disciplines" as it relates to AP Automation. Later in this section I will give you a workable test you can administer to ensure that each discipline will not be a negative factor in your world.

Discipline 1: Network Scalability

For starters, let's go easy. As explained in the Dictionary definition, network scalability is pretty easy to understand. To make the definition somewhat entertaining as well as

---

[4] http://dictionary.reference.com/browse/scalability

informative, I'm going to use a highway and cars as an illustration.

In Figure 2 the highway is open. Cars are moving at an appropriate speed, they can exit and enter and traffic crosses without a problem. This is just like a network. The road is the infrastructure and the cars are the data.

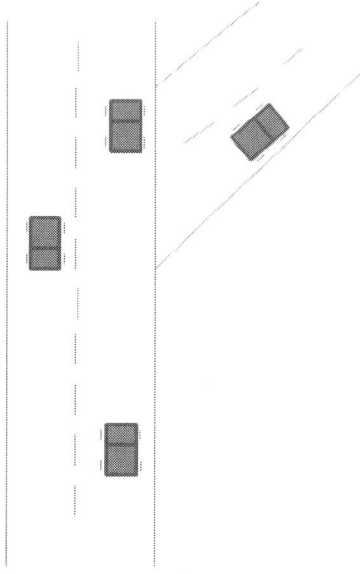

Figure 2

If the road stays the same and the traffic increases the cars (data) will slow down (Fig. 3). If the cars (data) continue to increase and the roads (network) do not, the cars (data) will slow down and sometimes come to a complete stop. (Right? – People in Atlanta or LA.)

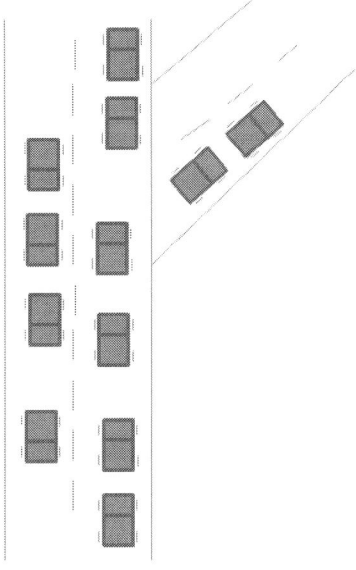

Figure 3

The problem is easily corrected when the road (network) is widened or more lanes are added (more bandwidth). This allows the cars (data) to travel without delay (Fig. 4).

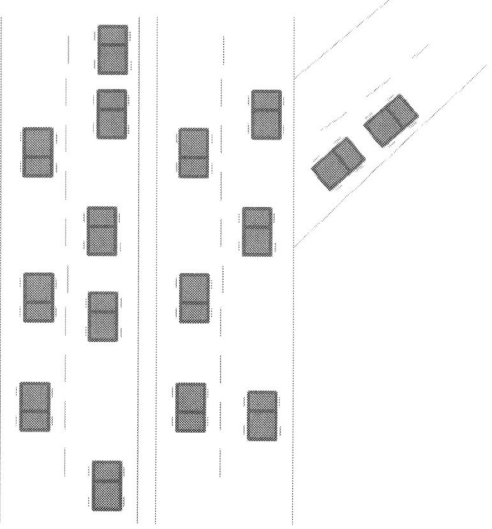

Figure 4

The widening of the road (bandwidth) is the addition of servers (RAM, Memory, Power, etc.) and a better connection to the internet (T1, T3 connection).

Discipline 2 – Code Scalability

With discipline 1, the situation becomes very frustrating when you add data, people, or load onto your network, and realize the network is not scalable. Add to the network (your servers or upgrade connection) just to find out that the speed is no better. This is where the second discipline and the most frustrating/hardest to correct comes into effect: Code Scalability, referring to the code within the software.

Using the same analogy of the road and the cars, the code is the engine that drives the car. The problem is twofold. First, the cars can break down causing a need to be fixed. A skilled mechanic (software developer) can trouble shoot and have the broken cars up and running. The second, and most difficult, problem to correct is cars driving in the wrong direction at the wrong time (Fig. 5).

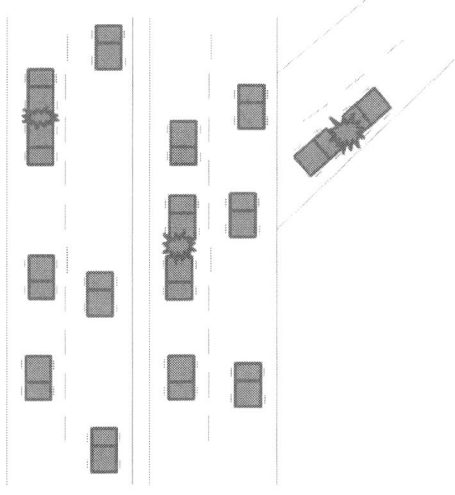

Figure 5

Fig. 1.4

Guess what will happen if cars on the freeway start heading northbound in a southbound lane. When you interact with a piece of software, it performs thousands of requests that create several actions per second. If the software is not programmed properly a request will get in the way of another request causing both requests to either wait on another request to sort the knot out or create an error that will not allow you to move on. If this happens, no network hardware or infrastructure additions are going to help code scalability.

Code scalability problems occur when a developer has no idea what they are doing or when a team is led by a person with little to no experience. To put it in more of a positive way; to eliminate code scalability problems the developer or leader needs to be very skilled on the platform in which they are building the software (road). It is critical for the developer to fully understand the goal of each process within the software. In the modern world, the developer has to have the ability to talk to people and interpret the problems they are trying to solve using software. Not wanting to put all the blame of communication on the developer, the person articulating the problem needs to have an understanding of the realm of possibilities within the software.

Developer(s) as well as the problem fixer(s) need(s) to be in concert on possibilities as well as delivery. If one or both are out of bounds on either the goal/outcome possibilities or developers abilities; serious problems will occur. Continuing with the idea of the road, let's say you have a talented construction worker that is skilled at pouring a driveway, not building a road using asphalt. That person is approached by a developer that wants to put in a new road to access another town in a more direct route; however that route is on a 90 degree grade. What do you think the outcome will be? It will not be good. Same with solving a process problem, if the goal is unrealistic and the builder (developer) is not skilled and experienced, the outcome will be an untraveled road with lots of bumps.

Discipline 3: Process Scalability

This is a difficult one too, because, unlike the two previous disciplines scalability is 100% made by humans. Sticking with the road, but expanding the idea, the cars are the software modules. A full software solution is made up of smaller pieces of software to manage particular processes and procedures. These smaller pieces of software are called modules (An example is within your email client you have the ability to search. The search function is a module). So, the modules are the cars, the road is the network and the process is the highway system. As you are aware our highway system is designed to take us from point A to point B in the quickest, safest, and most efficient manner. The same is true with the process that is outlined to automate accounts payable. Experience plays a key role in constructing the correct highway system. However, with process scalability the experience comes from (generally) non-technical sources, and the problems start because the sources can be (and usually are) numerous. The old saying, "Too many cooks in the kitchen" rings truer than ever. Here is a list of influences on the accounts payable process:

> *If the parts of the process are not outlined in the proper manner your highways will start creating circles*

- ➢ Auditor
    - o Internal
    - o External
    - o Customer
    - o Investor
- ➢ Approvals
    - o Budget (over or under)

- o Accuracy of the work
- o Product Delivered
- o Contractual
- o Against Purchase Order
- ➤ Coding
- ➤ Control
  - o Owner
  - o Department
- ➤ Accuracy of the Billing
- ➤ Vendor
- ➤ Risk
  - o Duplicate
  - o Non – Vendor
  - o Scam – Fraud

This is not a complete list; however it starts to make the point about the horrors of process scalability. If the parts of the process are not outlined in the proper manner your highways will start creating circles. Figure 6 illustrates the correct highway system, with exits and sections created in a constant motion pattern.

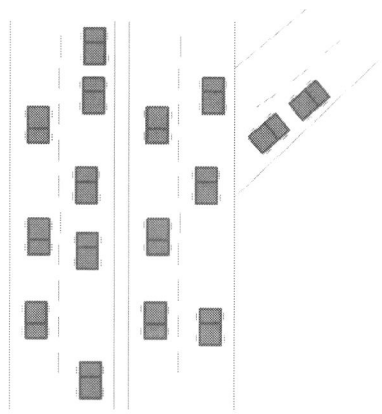

Figure 6

Figure 7 illustrates the dreaded principle of looping. Looping occurs when the same person needs to touch the same invoice multiple times in different steps of the process. Looping will kill the efficiencies of your process as well as eat up very large chucks of your processing time. This will lead to late payments, fire drills, and you can forget the ability to capture a vendor discount.

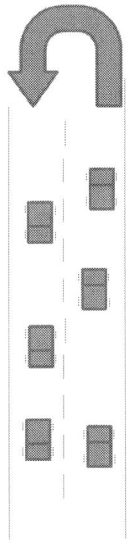

Figure 7

Understand that if there are loops in your process, then the process is, by definition, not scalable. Here is an example:

Approver 1, Greg, is an accounting person who will receive the invoice and code it against budget. Approver 2, Peter, is the approver of the invoice against receiving and the contract. Approver 3, Janet, is to approve against Sarbanes Oxley (SOX) compliance rules. After these three approvals the invoice needs to go back to Greg to be entered into the accounting system, then to Peter again for cash management. This process may work well when the invoice volume is low, but if you add additional volume the process will overwhelm due to loops and re-work (re-work is having to repeat the same work more than once. The difference between re-work and looping is re-work can be done by different people but it is the same work); therefore the process is not scalable.

> *The largest cost of an automation implementation is trial and error.*

So...now that you understand the disciplines of scalability: Network, Code and Process... why does this make a difference? The one word answer is money. The largest cost of an automation implementation is trial and error. The best way to combat trial and error is to create a system in the beginning that will grow with flexibility and not to bring you to a certain point and then require you to do something completely different.

Now it is time to tip the scale in your favor. Use the questions below to determine how scalable your (current or potential) automation process is.

Questions	Yes/No
1. Is your internet download and upload speed higher than 1000 kb? If you do not know, go to http://www.speakeasy.net/speedtest/ and find out.	
2. Did the developer(s) or company who created your automation software have more than six years' experience with creating AP Automation? Based on experience, three years is the	

8 PITFALLS OF ACCOUNTS PAYABLE AUTOMATION

First Edition 2011

break point that understanding of AP Automation because less theoretical.
3. Is the maintenance of your network done by a person where that is their sole job?
4. Do the developer(s) or company who created your automation software do AP Automation exclusively?
5. Did a process designer help you develop your process?
6. Was your process designed from scratch instead of making your paper environment digital?

Score: For every yes give yourself 1 point

Score	Scalability
6	Very Scalable – No Problems
4 or 5	Mostly Scalable with long term problems (over 10 years)
2 or 3	Somewhat Scalable with midterm problems (5 to 10 years)
0 or 1	Will not scale – Immediate Problems

(Note: for a printable version of the above chart/questions go to http://8pitfalls.com/ws/worksheet4.pdf)

The next time you have the privilege of participating in a software demonstration, go ahead and ask the sales person/engineer to step on the scale. It is perfectly acceptable to use the above disciplines to find out if their solution is going to help, hurt, or cost you in the long run.

Chapter 3

Getting Organized

Pitfall – Doing Your Homework Will Keep You Out Of The Pit.

Accountants and other members of the accounting department get a very bad rap. Often, the accounting department is labeled as not being flexible or having no ability to change. I believe this comes from their strong sense of responsibility towards the job of protecting the company's books (money). Some departments work for years to create a system that is free of holes where money can run and hide. When change comes from outside accounting and challenges the system created, the initial response from Accounting has been.... NO! Change for the sake of change is no good; it can threaten the accounting department's hard work in creating a system with no holes. However, the winds of change are here. With the advance in technology AP Automation is not just for the Fortune 1000 anymore.

Many small to medium-sized companies can now take advantage of web-based AP Automation without heavy upfront costs or set up. However, if the change is not done correctly, then "out with the old and in with the new" can bring pain, heartache and waste lots of time and money. If the change is done correctly, AP can become more efficient and can even become a strategic group within the organization. This chapter focuses on preparing you and your company for the change as well as maximizing your efforts to ensure that all the moving parts become connected.

Organizing Your Systems

When taking on an automation project; without first organizing the elements of the project, a company will create patches, fixes, add-ons or changes to the system to correct the process after it is started. All these unnecessary changes to the process or system waste time and money. The primary goal of this chapter is to keep you from needing to change or update your systems as you go along by giving you a roadmap for organizing your project upfront.

Another goal of this section is to make sure you do not duplicate your technology efforts. By not organizing first, you may end up creating a mixture of different systems with different goals. As mentioned in the introduction, this is not a technical book, so this chapter is not about computers, servers, or software mapping. It is about how to create a strong plan upfront to ultimately reach your automation vision.

Types of Projects

There are three types of AP Automation projects. Identifying the three types will help to show which project is best for your company along with other considerations once your project has been defined.

The three types of AP Automation projects are: Rehab, New Construction and Hybrid.

Rehab Project

The first type of project is simply taking your company's current process and making it electronic. I call this a Rehab Project. Generally, with this type of project the company has had a strong track record of getting invoices coded, approved and into the accounting system. There is normally a well-written procedures document with clear instructions that all employees adhere to as it pertains to the AP process. In my experience, in larger companies and/or public companies the need to follow the letter of the law is paramount. Large companies need written documentation and have strict rules about procedures because their size prohibits unspoken rules or one-on-one on the job training. Additionally, public companies have documented procedures due to government regulations (Sarbanes Oxley) that require the AP processes to be consistent.

Prior to a rehab project your company needs to make sure that the vendors, service, and software are able to mirror the already developed process. For example, let's say your current process requires dollar amount approvals for general operational expenses. If the service provider selling you their AP Automation software does not have conditional workflow rules by dollar amount, then you will need to look for another software service provider.

> *A paper-based AP process results in the need for multiple checks and balances to ensure that the paper invoices are not lost*

Be aware that a paper driven process naturally causes what I call "paper dysfunction". A paper-based AP process results in a need for multiple checks and balances to ensure the paper invoices are not lost, routed to the incorrect location, or purposely hidden by employees. Every company has a similar story (usually more than one) demonstrating how their paper-based system caused control problems. For example, I have worked extensively with Multi-family real estate companies who own and operate apartment buildings. These companies set up incentive packages for Property Managers to stay within the company's budget for each property. On occasion, a single invoice would push a Property Manager's budget into the red and, simultaneously, the invoice would somehow mysteriously disappear. The Property Manager had hidden invoices in his desk drawer and the invoices would only reappear when the Property Manager left the company and his desk was cleaned. Due to situations like this, Accounting departments create stop gaps, so they receive the invoice before the Property Manager. In an automated world, the extra step of Accounting receiving the invoice before the Property Manager would not be needed. The invoice would be received directly from the vendor and tracked electronically from the initial date of receipt.

New Construction

The second type of project, New Construction, typically creates an entirely new AP process. Just as the category name suggests, this type of project starts from the ground floor and builds an entirely new AP process. One reason companies head in this direction is that automation will bring such a significant change, it makes better sense to scrap the old and start fresh. Another reason is that no current documented procedures exist, so the change will present the perfect opportunity to create them. In my experience, when companies have chosen to scrap everything and create an entirely new process, the fresh new process (in the end) gives the AP department and its employee a brand new environment...something they are proud of, something that is cutting edge and adds value to the company as a whole. Be warned, however, this type of project, and its wonderful outcome, requires a very strong leader (or leaders) working together with the automation service provider to create the desired outcome. The leader(s) within the company must have autonomy along with a clear vision of the ultimate goal of the project.

One of the most difficult aspects of getting organized to automate AP is defining the vision. It is very difficult to have a vision for something that has never been done before. Within the company, the leader of the "new process" will go a long way with a "new construction" project if they have the backing of company as well as the trust of the employees in the accounting department. In other words, starting completely over is not an easy thing to do, it can also be the risky, but, if done correctly, the results are tremendous.

Hybrid Project

The last type of project is the hybrid approach. With a hybrid project, parts of the company's current process are working well, while other parts need to be reevaluated and/or completely changed. For example, the company's overall corporate AP process is running smoothly, but the legal department's approval process needs some work. All corporate invoices are paid in good time and the approval process is under control, so the company simply needs to digitize the invoices. However, on the legal side of the company invoices are often lost, resulting in late fees - the current paper approval process is creating more harm than good. A hybrid approach would digitize the corporate invoices (not changing the process), while scrapping the approval process for the legal department and starting over.

Exercise: Project Questionnaire

The following list of questions will help you decide which type of project is best for your company. Please answer Yes or No.

Questions	Yes/No
1. Are your procedures written	
2. Are the written procedures used? (Careful, be honest with yourself.)	
3. Have the procedures been updated within the last two years?	
4. Was an outside company (consulting or audit firm) used to write the procedures?	
5. Do the written procedures include levels of permissions (ex. coding an invoice, changing amounts, disputing an invoice)?	
6. Do the written procedures include exceptions (i.e. amount approval breaks, over budget)?	
7. Are all invoices received in a central location?	
8. Is there a committee dedicated to the administration of the procedures?	

For questions 1-7, add one point if you answered yes. Add two points for question 8, if your answer was yes.

Total all the points: _____

Key:

Score	
9	Rehab (Convert paper to electronic)
5-8	Hybrid
4 or less	New Construction (Complete overhaul)

(Note: for a printable version of the above chart/questions go to http://8pitfalls.com/ws/worksheet5.pdf)

Workflow:

Chances are, after the last exercise, your company falls into the Hybrid or New Construction projects. However, if you are clearly in the Rehab project section, read this section anyway... its brilliant! In the Hybrid and New Construction projects, there is a high need to define an approval process, called workflow. Workflow is a predetermined electronic approval process. This is often a confusing term because in the world of automation, workflow may be used in referring to the engine or software that drives the approval process, or the approval process itself. Within the workflow process, there are two other terms that are important and need to be defined. First is the concept of Role. Roles, similar to job titles, are groups of people that have a certain set of permissions and abilities. Roles determine the level of abilities and permissions, as well as authority within the workflow process. The other term that I will use throughout this book is Approver. Approver refers to the person doing the work within the workflow process. In this section, I will outline how to create and map your approval process, give you pointers on improving the process, and provide a list of things to stay away from when creating workflows.

Creating the process

I mentioned earlier that no two companies are structured the same way as it pertains to the AP process. Yet, I have also found that most companies are identical when it comes to the types of invoices that need to be approved. Examples of invoice types include: marketing, legal, office, technology. To create workflow, the best place to start is with the types of invoices. A bit of advice: during this exercise - don't worry about how the automation software or service works, just focus on the types of invoices first, then identify how to modify the software or service to fit your needs.

Here are the steps to identifying and defining your workflow processes:

1. List all of the departments within your company, for example, marketing or legal. (Note: Write this on the left side of the paper and leave space between each department.)
2. List the current invoice approval process within each department. (Note: Be specific and include names and job titles.)
3. For each job title, list the approver's permission levels such as coding the invoice, changing amounts, disputing invoices, etc.
4. Based on the department approval process, list the possible exceptions to each approval. (Note: Be specific and include names and job titles.)

Your list might look something like this:

Department	Approval Process	Permission Levels	Exceptions
Marketing	John (Marketing Assistant)	Code, review for reasonableness and approve	
	Sally (Marketing Manager)	Review and approve	If it is a trade show go to Peter
	Peter(VP	If over $5000 or a	

		Marketing) if over $5000	trade show review for reasonableness and approve	
		Accounting	Review codes and approve	
	Legal – Internal	Jan (Paralegal)	Code, review for reasonableness and approve	
		Accounting	Review codes and approve	
	Legal – External	Bobby (Legal Admin)	Code, review for reasonableness and approve	
		Mike (CFO)	Review and approve	If client is involved and over $25000 go to John (CEO)
		John (CEO) If over $25000	If client is involved and over $25000 go to John (CEO)	
		Accounting	Review codes and approve	
	Office	Receptionist	Opens mail, sorts and places in appropriate person's mail slot	
		Assistant Manager for the department that ordered	Codes and approves	
		Mike (CFO) if not on budget or over $5000	Reviews and approves capital expenses over $5000	
		Manager of Department that ordered	Reviews for informational purposes	

		Accounting	Entered into AP System for payment	
███████████████████████████████████████
IT – Network		Cindy (Network Admin)	Code, review for reasonableness and approve	
		If on contract John (Director of IT)	Review codes and approve	
		If off contract Mike (CFO)	Review codes and approve	
		Accounting	Review codes and approve	
███████████████████████████████████████
IT – Software		Alice (Software Admin)	Code, review for reasonableness and approve	
		Accounting	Review codes and approve	

With the current state mapped, underline or highlight each step that is not an approval step. A good definition of an approval step is any person that needs to approve the goods or services that have been received, approve the coding, approve against budget and/or approve the dollar amount. You need to eliminate any person or step that is not critical to the approval process. Having too many people in the approval process will lengthen your cycle time and unnecessarily cause your payments to vendors to be delayed. Cycle time is measured from the time the invoice is received to the time the invoice is fully coded, approved and entered into the accounting system. The longer the cycle time, the

> *You need to eliminate any person or step that is not critical to the approval process*

more chance the invoice has to get lost and the higher chance you have of incurring late fees. Also, the longer your cycle time, the more expensive your approval process is. In an automated approval process, people who only have the need to view the data (not approve it) can be satisfied by sending that person a report with the information they need rather than making them a step in the approval process. Cutting out approval steps increases cycle time and streamlines your approval processes. Here is an example using the Office invoices workflow previously listed:

Current State:

Office	Jane (Receptionist)	Opens mail, sorts and places in appropriate person's mail slot
	Assistant Manager for the department that ordered	Codes and approves
	Mike (CFO) if not on budget or over $5000	Reviews and approves capital expenses over $5000
	Manager of Department that ordered	Reviews for informational purposes
	Accounting	Entered into AP System for payment

Approvals Only:

Office	~~Jane (Receptionist)~~	~~Opens mail, sorts and places in appropriate person's mail slot~~
	Assistant Manager for the department that ordered	Codes and approves
	Mike (CFO) if not on budget or over $5000	Reviews and approves capital expenses over $5000
	~~Manager of Department that ordered~~	~~Reviews for informational purposes~~
	~~Accounting~~	~~Entered into AP System for payment~~

In the example listed above, notice the three steps eliminated due to automation. First, the invoices do not need to be received, opened, or sorted. In a fully automated process, invoices will be received electronically and sent to the AP Automation system electronically. (For more information see Chapter 5) Secondly, the Department Manager wanting to be in the loop will receive a report that outlines all the costs related to their specific requirements. Lastly, the invoices do not need to be manually sent to the AP department for payment. They will be electronically passed to the accounting system from the AP software or service once they are fully approved.

In the example above, the fourth step of the approval process is a Department Manager who wants to monitor the expense. Only people who approve invoices need to be in the workflow, everyone else should receive a report. This is especially true for invoices that are time sensitive; it never serves the process well to have too many people in the middle. However, there may be a very good reason a non-approver is in the middle of the process. You may simply need to adjust the automated process to better fit the needs of the company. For example, utility invoices are on short leashes, meaning the time the utility provider gives your company to pay the invoice could be only 15 or 20 days from the date of the invoice. If the terms aren't met, the utility will assess a late fee on the next invoice. If the late fees are not paid, they turn off the lights (they have you right where they want you). Due to the tight timeframes, accounting needs to get the utility invoice into the accounting system and paid quickly. However, the person or department responsible for making sure the charge is correct is not inside the accounting department; therefore, the invoice needs to go to the person or department responsible for the charge before being entered into the accounting system. This is a potential time killer and cost buster. In an automated world, the invoice would go directly to accounting for coding and approval, then to accounting system for payment. Once the approval has been made, a report can be generated each week or month and sent to the person or department responsible for the charge. Using this process, the invoice gets paid on time, incurring no late fees. The responsible person or department can argue any charges after the payment is made and have it credited against the invoice.

A Few More Things to Consider

When crafting or evaluating workflow, make sure the steps in the workflow always lead in the direction of the accounting system. Situations can arise in which an invoice is at the end of the approval process and is still incorrect. The person who caught the mistake at the end of the process would typically send the invoice back to the person who made the mistake. This is called looping or circles in your approval process. Looping typically occurs when the permissions are not properly mapped, as shown in Figure 8 below. For example, some companies consider sending the invoices from accounting at the end of the process back to the person who made the mistake. This type of loop will destroy your invoice process cycle time.

Asst. Mgr → Manager → Accounting

- Code
- Vendor

- Edit
- Change $

- Edit
- Reassign Workflow

Figure 8

In Figure 3.1, the approval process begins with the Assistant Manager and ends in the Accounting Department. Listed below the titles of each step are the permissions of each approver. All permissions need to work in sequence together and lead the invoice toward the accounting system. The problem in Figure 3.1 is, if Accounting received the invoice and tagged it to the wrong vendor; they would have to send it back to the Assistant Manager. As another example, if the Assistant Manager received an invoice but the invoice was originally placed in the wrong workflow, the invoice would need to be sent to Accounting and reassigned. Design your workflows so that the invoices are always heading in the direction of the accounting system and be mindful to eliminate any "loops" along the way.

Empower Decision Makers

One of the most difficult principles about automation comes around the idea that workflow needs to empower your decision makers and not take the place of decision makers. People should, not software, make the decision in workflow about invoices. Here are two reasons why:

1. People think more critically and creatively than software
2. People are ultimately responsible for the expense.

People Can Think

I always have trouble explaining this aspect of automation. It comes from a misunderstanding. Do not get me wrong! Automation and workflow are very powerful. However, automation, workflow, software, and technology cannot think or react like a human (not yet, at least). I am very sorry for creating this controversy among my Sci-fi friends. One day (maybe soon) I will have to revise this section as the abilities of software become greater, but for now a realistic goal for AP Automation is to empower your employees to do the critical work. Here is an example of critical work: An invoice comes in and it is not attached to a purchase order, so it needs to go through a departmental approval process. When the invoice is assigned a workflow, (electronic approval process) the workflow steps are: (1) assistant department manager for coding, (2) department manager for budget approval, and (3) accountant for cash approval. Let's look at the possible options for how people and technology look at the above scenario.

Step	Technology	Person	Conclusion
Step 1 - Coding	Technology will have the opportunity to do a lot of things in this step. A vendor can be associated to a default General Ledger code or an allocation. Some software has the ability to recognize patterns in the users coding habits and assign those patterns.	People have the ability to read beyond the needed black and white patterns of technology. An example is a landscaping invoice. Technology can assign default coding, but a person can read that a particular invoice is for snow removal which does not fall under the default code.	In coding, technology is extremely helpful and may be able to "pre-code" a certain percentage of invoices, but ultimately a person will need to make the final judgment on the correct code.
Step 2 - Budget	Once the invoice is coded, technology has the ability to look at the coding structure as well	Very similar to the coding step, a budget step can be greatly aided by pre-populated information, such	Budgets can be complicated. Multiple codes, time periods, and dollar amounts can be

8 PITFALLS OF ACCOUNTS PAYABLE AUTOMATION

First Edition 2011

	as the invoice or accounting date and check a budget table for funds. This process can enable the technology to react to the conditions of the budget and stop or change the workflow.	as amount budgeted for a month. Unlike coding, the budget step relies on technology to do the math on budget by period both on what has been used and what has been committed. However, a person looking at the invoice can tell if the items on the invoice are non-budgeted items or items that are somehow "out of scope" to a project or for a particular month. It would be impossible for technology to make such an observation.	considered. Given all of the possible options, technology is extremely useful when displaying budgets. If the budget is overdrawn, workflow can be helpful in routing the invoice to additional approvals. The overall conclusion on budgets is it works well until there is a judgment call on the expenses that are outside of the budget but need to be coded to a budgeted line item.
Step 3 - Cash	For cash, technology is able to tell you how much you have and where the money has gone. Technology can also provide a person with alerts on certain factors, such as an account getting low or a certain dollar amount has been exceeded.	As well as technology can track cash, ultimately, a person needs to be responsible for the money of the company. It is just too important to leave the management of cash to technology alone.	Cash management is too important to leave to technology alone.

The above scenario is designed to illustrate how technology and people can work together. Workflow does a fantastic job of getting the invoice to the right place at the right time for a person to make the right decision.

People are Ultimately Responsible

If you think about it, accountants are hired to guard a company's assets. Part of guarding the company's assets is ensuring that invoices are coded properly and entered into the accounting system correctly. Ultimately, the employees within the accounting department are responsible for each of these actions. It is unlikely, anytime soon, a piece of technology can fully replace this responsibility.

> *The AP staff of 7 people was spending 80% of their time on 10% of their invoices.*

To demonstrate this idea of enabling your workflow to empower your decision makers let me tell you about a large manufacturing company I worked with in Georgia. They processed over 250,000 invoices per month. The majority of the 250,000 invoices were being matched to a purchase order and approved in their ERP system. That part of the process was not the problem. The problem was with all of the invoices that did have a PO where the PO and invoice did not match. From an overall perspective that represented about 10% of their invoices. Meaning 25,000 invoices were being routed and approved as paper. To clarify the problem, the AP staff of 7 people was spending 80% of their time on 10% of their invoices. It was the exception invoices that were eating up all of AP's time. The company desired to grow, but the accounting department was unable to keep up with current demand, and was terrified of the idea that growth was just around the corner. At the beginning of our engagement it became clear to me what the AP department had in mind to fix their problem... OCR. OCR stands for Optical Character Recognition (see chapter 5 for more information about OCR). It was believed (by the AP staff) that OCR would read the invoice; look for key terms and, based on a set of mathematical formulas, approve the invoice, therefore achieving the AP department's ultimate goal of not letting the invoice out of their sight. I told them it was not a good idea and the OCR engine would not be able to do the critical thinking like a human. They said, "Well I guess you are not our guy" (for the record, they said it in a very nice way). My concern for them is they would be replacing one problem that consumed their time with another problem that consumed their time (maybe even more of their time), which is exactly what happened. Their original problem of passing paper around for approval worsened because of a new problem of handling all the exceptions the OCR engine produced when it was unable to think outside of the (computer) box like a human. Let your software manage the process and empower your people to make the decisions.

Other Considerations:

When creating or updating workflows you should also consider the use of contracts, purchase orders (POs), and purchase cards (P-Cards).

P-Cards are helpful in speeding up the workflow process because they reduce the number of invoices that need to be managed. P-Cards work just like a credit card, but have embedded rules setup by the company so the card can be used with preapproved purchasing limits per user by types of purchases. Typically, a company will use a P-Card for items or services that are under a certain dollar amount, for example $100. Any purchase under $100 is charged to the card, paid and evaluated outside the workflow process. Because one invoice from the P-Card vendor can handle many types of purchases as well as multiple vendors, the numbers of invoices that are processed are reduced. For example, a non P-Card organization may currently process 3000 invoices a month. If the same organization used a P-Card for all purchases under $100, the invoices managed in the automated workflow process would drop from 3000/month to 2500/month. And therefore the burden on the workflow would be cut by 16%.

Contracts and Purchase Orders (PO) have different purposes and offer unique control advantage. They have a similar impact on the workflow process to P-Cards. Contracts are negotiated ahead of time with vendors and expenses related to the contract are tracked over the course of the contract. On the other hand, POs are used to get approvals in advance to purchase something and are normally a one-time situation. Because both contracts and POs obtain prior approval for the purchase, if used properly, contracts and POs will take some of the burden off the invoice workflow process by eliminating approval steps.

Here is an example of using a contract. An optimal workflow process begins with the Assistant Manager, travels through the Manager and Regional Manager to Accounting. If a contract is in place with that vendor and an invoice amount/quantity is within the terms of the contract, the workflow process could look something like this; Assistant Manager to the Manager (as long as it is on contract) and then on to Accounting. The Regional Manager is bypassed because he/she originally negotiated the contract and the invoice falls within the terms he has already indirectly approved.

> A well-crafted workflow will solve the problem of late fees, lost invoices, slow approvals, numerous staff members wasting time, slogging through the monthly accrual process

In a 3-way PO matching scenario, a) the PO is created, coded and approved, b) the goods are received or the services completed and c) the invoice is compared to the original PO and approved. If the automated system matches the invoice dollar amount to the PO, the only human intervention in the process is the receipt of goods or the verification of the completion of service as the invoice goes to Accounting. The key here is the invoice was pre-approved based on the information in the PO and as a result cut out workflow steps in the process.

Looking back to earlier in this chapter, I explained workflow can be slowed or encumbered by creating too many exceptions, or not having been configured correctly because of incorrect assignment of permissions. I also introduced the idea that only people who approve invoices need to be in the workflow steps and everyone else can receive a report. So let me pull this all together and show you an example of an optimal workflow process. Notice in Figure 9 below that the permissions build from the Assistant Manager to the Accounting person. Each step is building on a hierarchy of abilities. This will allow for the invoices to continue on through the process getting the correct approval without having to loop, or reverting back to people earlier in the workflow. Also, reports are triggered by events and not people, allowing individuals within the company to monitor expenses without holding up the invoice's approval progress. Lastly, conditions are embedded in the workflow, such as the over $2,000 step to the Senior Vice President, which allows for faster approval of invoices and reduced the number of invoices approved by the Senior VP. When crafting an optimal workflow, speed and accuracy need to be a critical consideration. A well-crafted workflow will solve the problem of late fees, lost invoices, slow approvals, numerous staff members wasting time, and slogging through the monthly accrual process.

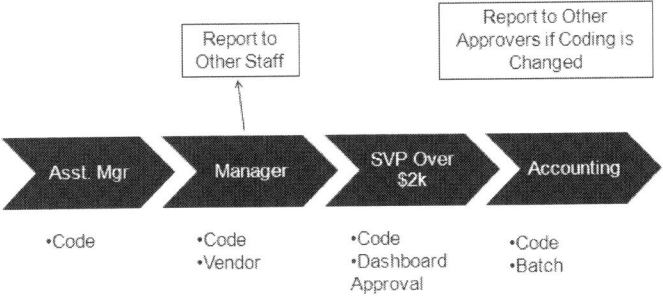

Figure 9

In summary, take the time to get organized before you start the project or physical work on automating AP. The "Trial and error" approach will make things difficult and waste a lot of time and money. Make sure to focus first on the process you are trying to implement and not on how the software works. Then, before you enter into the project, determine what type of project you are going to do (Rehab, New Construction, or Hybrid), and clearly map your approval paths and permission levels. Following these steps will allow your company to go smoothly into the implementation of your new AP Automation tools on time and within budget.

Chapter 4

The Project

Pitfall – Perception Is Not Reality. Keeping The Project Organized And In Control Will Keep You Out Of The Pit.

The majority of my experience has been managing AP Automation projects. I have also been the person who created the methodology to ensure success. However, even though this chapter is dedicated to help you stay in control, it is practically impossible to outline a project pattern that works 100% of the time for every company. This chapter will give you four questions to ask as well as the six major steps in completing a project.

Remember the System of Record

Your company's accounting system is always the system of record. Meaning whenever financial reports are run, checks are cut, or cash is managed, it will always be in the accounting system. Beware! There are automation companies out there that will compete with your accounting system for dominance. Here are the four questions to ask in order to determine if there is going to be competition.

Questions:

1. Does the AP Automation system update GL directly?
2. Will the AP Automation system create a vendor with an ID?
3. Does the AP Automation system help manage cash and bank accounts?
4. Is the integration of the AP Automation system custom?

Answering yes to any of the above questions will indicate there is going to be competition between the AP Automation system and the accounting system. When writing competition, I am referring to which system is going to make critical financial decisions. Chances are, unless you are running your accounting on spreadsheets, your accounting system has an Accounts Payable module. This module is designed to take approved expenses and post them to the general ledger to be paid. When automating AP the process/service/software you will need to work with that AP module, not replace it. The reason for this is to maintain secure cash management. If you have more than one system of record you will greatly increase the chances of invoices being double approved, not approved at all or not correctly approved. Based on the questions above, the key to look for is where all the heaving lifting is done. Where are the vendor records being managed? What system has access to the General Ledger (GL)? To ensure the system you create is fully secure make the account system the system of record.

What Type of Project is the Best?

Once you have clearly established the accounting system as the system of record, the big question that will determine the type of project you need is: Is there a pre-established integration to the accounting system's AP module and the AP Automation software? If the answer to this question is no, you have your work cut out for you. There are two types of integration, synchronization (which is a free flow of information back and forth from the AP Automation software to the accounting system) or an upload-download process from the AP Automation software to the accounting system. The thing to look for when building the integration is error checking and error handling, because getting the data into the system is not the difficult part. Getting data into the system in a secure manner is the difficult part. When a payables file moves from the AP Automation system to the accounting system items such as vendor/department/cost center/location/business unit/entity need to be checked to ensure they have not been deactivated. Other items such as accounting period, code associations (i.e. department to chart or GL codes) need to be verified. Notice, however, I identified not only error checking, but error handling as a possible problem. It is not good enough to designate the error as an error; the error needs to be displayed in a usable form. I will give you an example. I worked with an accounting system that built a real-time custom integration with extremely clear error handling except for construction accounting invoices. If there was a problem with the construction codes, the errors would show in a SQL script with an *Ok* or *Cancel* button. If you clicked *Ok*, it would allow the corrupted data to load into the accounting system and cause trouble for the accountants. Also, the people pulling over the batches did not know how to read SQL scripts.

Point number one is to pick a company that has a preexisting integration/partnership with your company's accounting system.

Point number two is make sure the company that is running the project, whether that is a consulting firm, internal staff and/or the AP Automation vendor, has several years' experience with automating AP. Automating AP is not that tricky, however, if the implementer's experience level is low you will be slowed with re-work and learning curves.

Point three is to confirm your project plan includes testing. Every AP Automation project I've been involved with the client ask me, what is the difference between those projects that succeed versus those that fail. The answer,

100% of the time, is testing. Testing does two important things for your automation project. (1) It allows you to make sure everything you scoped and built works. (2) (This is the good one) Testing provides built in training time in a safe environment. This will allow you to train a few of your users to become super users. In the projects I managed, I asked the company to identify those people that would be system administrators. When I wrote the testing scripts I trained the system administrator on the testing procedures and have him/her do all the work. At the end of the testing period, the system administrator had the process and trouble shooting down, as well as had developed all of the advanced administrative skills needed. It was a win – win – win. A win for my company because there were onsite experts, a win for the client because administration work could be done within the company (there was no longer a need to call a help desk therefore things moved faster and more accurate), and a win for the administrator because he/she was never put in front of the fire squad ill equipped.

Point four is to plan and execute a phased rollout. This final point has the most debate, so here is my reasoning. A phased role out consists of adding location by location or department by department until the entire company's payables are automated. Because a lot of my experience is in Real Estate, it was natural to do a pilot property, then 30 days later add another property and then 30 days after that (now 60 days into it) add the remaining properties. This allowed me and my client to make sure there were no procedures or exceptions missing. For those companies that did not do a phased rollout (I called these slams)... they set themselves up for a potential disaster. The disaster was not being prepared for the unknown.

Notice, however, I am suggesting rolling out a few locations or departments and then rolling out the rest all at once. The reason for this is to find those things you may have missed in a low risk environment. A good way to think about this is to separate the "going live" into two sections. There is the implementation or the configuration of the software (I like to call this the build), and there is the processing and receiving of the invoices (I like to call this the switch over). I highly recommend that you do the "build" portion all at once. I also recommend that the "switch over" be more controlled and slower over time. This way you are able to concentrate time on the build and not have to do dozens of additions to the configuration of the automation software. Also with the slow, controlled switch over you can ensure that nothing was missed in the build.

6 Steps to Greatness

It is important to use a project methodology for automating AP. Consulting firms will most likely disagree with this next statement, but I believe all project methodologies are pretty much the same. For example:

Figure 10

8 PITFALLS OF ACCOUNTS PAYABLE AUTOMATION

First Edition 2011

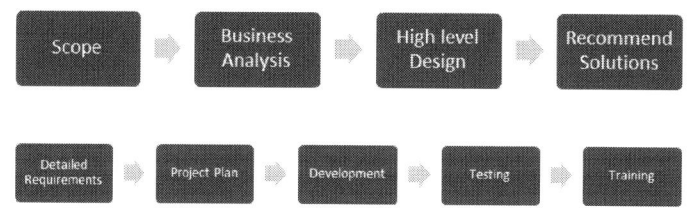

Figure 11

The broad steps to a project are simple. Each step is described below. However, I am not going into detail of each step in an attempt to stay focused on AP Automation and not project management. A good resource to get more information about project management is Project Management Institution (www.pmi.org).

1. Scope - Current Environment + Completion Goals
2. Build - Configure or Install
3. Test - Examine all process and connection
4. Train - Teach users the new process and software
5. Go Live - Completion date
6. Evaluate - How did you do, what would you improve?

Scope:

Scoping a project is the process of making sure your overall goal is supported by smaller goals. The difficult aspect in scoping is that each project is different. Factors that can change a scope in a project are (but are not limited to), the software supplier's support structure, the type of software (i.e. ASP, SaaS or client server), the company's internal skill set, and basic business timing. In order for the scope to be clear, usable, and achievable each goal needs to be outlined in detail, timed and assigned to the proper staff with dates on deliverables and completion.

Build:

The build phase of the project is just that, physically building the project. Physical work consists of installing the software on a server, creating the proper firewall procedures, installing local software on machines, and configuring the software. If you decide to use Software as a Service (SaaS) modeled solution, then the build portion of the project concentrates more on configuring the software for your company's unique use. The build phases should be tightly timed and the expenses tracked to ensure the project stays on budget.

Test:

After everything is built, everything has to be tested. Testing can be as simple as running a few invoices through the system and loading the approved invoices into the accounting system, then deleting. Or testing can be as complicated as having multiple test environments running dual processes on automated scripts. The purpose of testing is to make sure the ideas scoped were built correctly.

Train:

This step is the simple process of training end users and administrators to use the software and learn the new process created through automation. This step is very important. So much so that I dedicated chapter six to the process of training and going live.

Go Live:

Going live is more of a benchmark than a block of time in which actions need to take place. The Go Live benchmark represents a day that everything has to be ready. Typically, in the projects I have managed, we would start with the Go Live date and work back to ensure we had enough time to get everything done. Even though Go Live is a set date, the rollout plan can (should) be given time. With a rollout plan, all the users will not be onboard by the Go Live day. Still, it is critical to have a single deadline date in order for the project to remain on track and be completed on time.

Evaluate:

The evaluation step has two sets of questions to be asked by the project. First, once the project goes live the project team needs to go back to the first step (scope) and ask themselves, "Did we accomplish all the goals we set out to accomplish?" To give you a little insight, the answer will be clear and is usually yes. However, the direction or steps that are needed to complete the project are not going to be as smooth as originally planned. With every project, tweaks to the scope will be made along the way. The tweaks, adjustments, and lessons learned are normal. During the project, you will be presented with roadblocks and obstacles that were unknown at the beginning; therefore the path from scope to Go Live will look different during the evaluation.

The second set of questions to be asked by the project team during the evaluation step is, "What did we learn throughout this project implementation?" What changes or improvements to the project, software, people, etc. should I know going forward? These are important questions to ask the team, because AP Automation is not an event, but a process and its improvement needs to be ongoing.

Final Thoughts:

It is better to use project methodology throughout all aspects of your project to ensure its completion on time and on budget. To recap, following the four points (keeping your accounting system the system of record, experience, testing, and phased rollout) will keep your project under control and out of the pit. There is nothing worse in business than the thrill of a new venture going horribly wrong due to faulty methodology. In managing projects, I always tell my clients that we were going to do as much planning as we can and then plan some more because it will free us up in later steps to manage the unexpected ...and believe me there will be plenty of the unexpected!

Chapter 5

The Vendors

Pitfall – Poor Or Misdirected Communication Can Throw You In The Pit!

8 PITFALLS OF ACCOUNTS PAYABLE AUTOMATION

First Edition 2011

The very first position I held with an AP Automation company was "Director of Supplier Enablement". The goal of the position was to contact suppliers to let them know we had tons (slight exaggeration) of buyers online waiting for their products. Over a couple of years an interesting things happened. The suppliers I talked with, which were quite a few, initially love the idea. I was calling on sales people at these suppliers; approaching them with an untapped market of new business. All I asked was margin, meaning my company would take a percentage of all transactions in order for us not to charge the buyers. To be clear, this approach to AP Automation was more from the Purchase Order (PO) and Procurement side. The invoice was an extension of the PO, so the initial response was positive. As the supplier dug into the opportunity it became clear to them that my company's software commoditized their offering. Two things happen when you commoditize a suppliers offering; (1) It erodes margin, putting the low price providers in a better light (2) It strips the supplier's ability to serve their client's needs. You would think a light bulb is a light bulb. Meaning if supplier A sold a GE FLE15 HT 3/2/827 for $5.50 and supplier B sold the exact same lamp for $5.25... supplier B is better, right? Not quite, even if supplier B saves you $.25 per light their shipping may be 20% higher with a minimum order and it takes 4 - 6 weeks for delivery. The other consideration was supplier A had a local sales rep that is available to meet you at your office, upon the visit says you can use the GE FLE15 HT 3/2/827, but have you thought about using the FLE26 HT 3/2/827, even though it cost more it will save your company 55% over 2 years. It became clear that those conversations were critical to my client's business, therefore the attraction of many suppliers and many buyers purchasing goods online became less and less. In the end it helped my company move more into the automated accounts payable business rather than purchasing. In the transition, however, we came to a greater understanding that the supplier is a key component and an automated offering had to include benefit for them too. This chapter will help you understand what benefit you can provide the supplier, how to ease their fear of automating, and lastly (because of benefiting the supplier) how you are able to make AP a profit center or a revenue generating entity.

Which is the Best

With technology there are always a lot of assumptions, especially with emerging technology (technology that is still growing and changing at a rapid pace). A lot of time these assumptions come from the creator or seller of the technology, therefore it may or may not be rooted in fact. An example of this is how to receive invoices. The question is what is the best method? Here is a short list:

- EDI - XML
- PDF – GIF – JPEG
- Email
- Paper
- Fax
- Text (tab delimited) or CSV
- Private Carrier (FedEx, UPS...)

There are others... but the above list is some of the basics. The above methods are divided into three types; Data, Electronic Form and Paper (We will talk more about the types later.) There are also a mix of transmission methods and formats. Let's take a few of these as examples to answer the questions of this section: What is the best method to transmit an invoice?

EDI is a direct feed of data from computer to computer. XML is a direct feed of data from software to software. A PDF file is converted from paper to an electronic form, and paper... well is paper. From the supplier's perspective, depending on their size, paper may be the best option. Lately email has become more and more popular with suppliers. So which is the best method... email vs. fax or paper vs. data? The answer is yes! Giving suppliers options is the best method, allowing them to self-select takes the burden off of your company's AP process to mandate change and allows your company the opportunity over time to train and change those suppliers that can be changed without creating a science project for the internal staff. Here is an example. Let's say an accounts payable department has 20,000 active vendors. Faced with the need to automate, you drive your vendor to a portal where they select from PO Box, email or data upload. Once all of your vendors select a method the number may look like this 40% PO Box, 55% Email and 5% Data. Based on those numbers a realistic plan for the next 12 months is to convert 20% of your PO Box (paper) to email and 25% of your email to data. The new numbers will look like this: 20% PO Box, 50% Email and 30% data (see graph). The best practice is to continue moving the percentage towards data. Data will provide your company with the fastest and most accurate response from your suppliers. In order to execute on such a plan you need automation software/service that is capable to allow options.

8 PITFALLS OF ACCOUNTS PAYABLE AUTOMATION

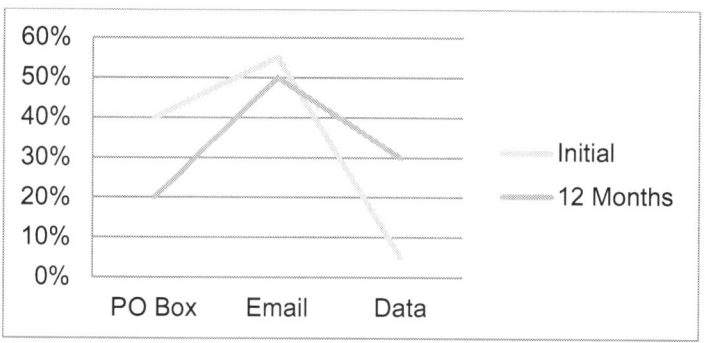

Figure 12

My Uncle Bought a Betamax

My Uncle Buddy was one of those guys that needed to have the newest, coolest and hippest stuff, and was a lot of fun going to his house to check out all the stuff. I remember him letting my Dad know that he now had the ability to watch movies at his home and could pause, stop and rewind (the good parts) as much as he wanted. This was way too much temptation for my family so we made the four hour trip to the big city to see this magical machine. It was called a Betamax. Buddy said it was the future of entertainment and everyone should buy one while the price was still a manageable $1200. In the mid-70s this was a lot of money. At my house we still had to hit the side of the TV when the picture that we had dialed in from a rotating antenna started looking like a zig-zag. Nowadays the plight of the Betamax is well known. (For those of you that do not know the plight of the Betamax, that should be a good indicator of its use and popularity). Looking back people should have never invested in anything with "Beta" in the title (that's a little software joke). However then – the Betamax was king. My uncle didn't own any movies but he had the machine.

A few months later a Japanese company invented the VHS and the two formats started fighting it out. It was a fierce battle that ultimately ended with my Uncle Buddy owning both at a large investment stacked on top of each other atop his TV (by the way – he bought Pong for his kids – that was awesome too). What made the feud so intense was that you couldn't play one format in the other machine. There was no crossover; you either had a Betamax with tapes or VHS with tapes.

Fast forward to today and look at invoicing. Ask a vendor if they can provide you with electronic data. Chances are they will say "yes". Then ask if they can send you a sample. Chances are they will email you a Microsoft Word document that is saved in Adobe's PDF format. This is not electronic data. Far from it! This seemly innocent slip up by your vendor is in reality a huge blunder.

Starting with the idea that your company wants to reduce cost by creating an Electronic Invoice Presentment and Payment (EIPP) process. The Aberdeen Group said that in 2008 the top consideration (at 41% of all of those polled) was to create a EIPP process as a key to automating AP (see Fig. 12) If you ask your top 10 vendors how many can provide electronic invoices, 9 out of the 10 will say yes. In reality, 7 out of the 9 can provide invoices in an electronic from, 1 out of the 9 can support EDI and the last one uses a third party content company that distributes its data though CSV.

In this section, we will go through what an electronic form is, what electronic data is and how the two are different. Not wanting to leave you out in the electronic cold, I am going to explain how both can be used successfully in an electronic invoice initiative.

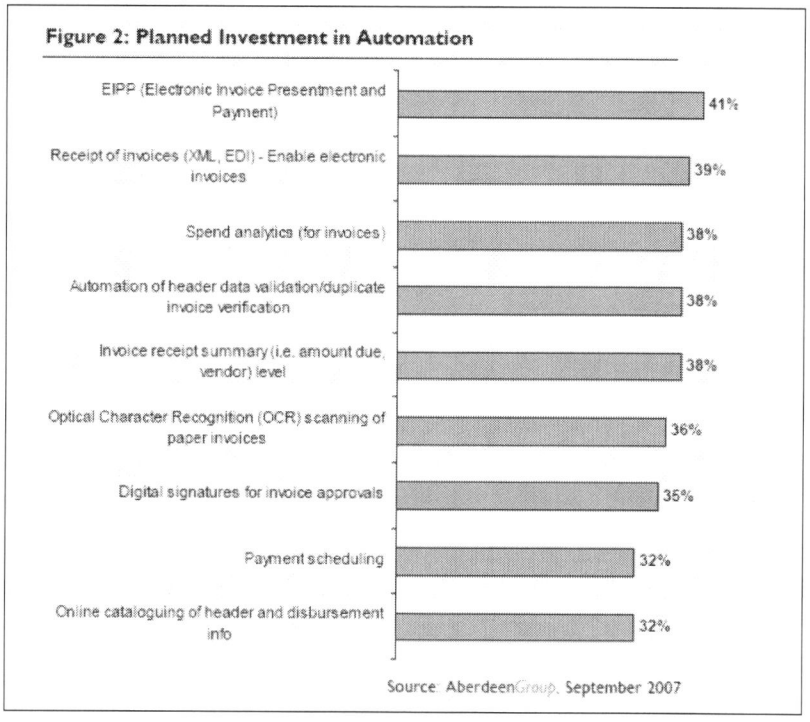

Figure 13

Different Shapes and Sizes, all One Form:

What is an electronic form and how does something like a paper invoice get into an electronic form? There are two ways. First, and most commonly, the invoice arrives to your location. Once the invoice is opened, it is scanned and saved as a TIFF, GIF or PDF (there are other types but these are the most common). Now, the scanned piece of paper is in an electronic form that can be indexed (which is typed into a computerized form) or saved into a workflow approval system, document management system, a shared drive, FTP site or accounting system. The scanned invoice is available in one of those systems to be accessed at a later time by more than one person. The scanned invoice is also available to be emailed or printed at another location. Here is the key to an invoice that has been scanned and is in an electronic from; you can shred or destroy the invoice. The electronic form is as legal as the piece of paper in the envelope. (Federal Rules of Evidence, Rule 1001<3>) The IRS also weighed in on the storing of electronic forms with the producing of IRS Revenue Procedure 97-22 (found in Internal Revenue Bulletin 1997-13.) Sparing you the pure joy of reading all 36 pages of the bulletin, the procedure redefines the term "document" to include "records maintained in an electronic storage system that complies with the requirements of this procedure (97-22). The requirements that the IRS is referring to is that the storage system housing the document has to have reasonable controls to ensure integrity, accuracy, and reliability, as well as the prevention of unauthorized additions, alteration and deletions, and is available for inspection. Other requirements are the ability of retrieval and the ability to reproduce legible and readable hardcopies (ref: Section 4 Sub Section .01 Line 2 a,b,c,d and e). So what this means in normal person speak is that the system storing the document needs to be password protected, with multiple levels of internal security to hide certain data from certain people with a recording mechanism that notes what has been done to each document. Last but not least, the system has to have the ability to print the document in a way that it can be read. So start scanning and saving to a secured location then destroy the original document and remove the filing

> *The electronic form is as legal as the piece of paper in the envelop*

cabinet to make more room for a ping pong table (or other cubes).

In my position as a consultant one of my main duties is to explain how electronic invoices are created and how workflows get assigned to invoices through the approval process. I dislike this type of question. The intelligence of the software(s) and the possible combinations of electronic feeds and forms for assignment to workflow are so numerous, it is impossible to explain in just a few sentences. Now when asked I just say, "It's magic". When it comes to technology, especially new technology, like electronic invoices saying "It's magic" is not too far from the truth. However, because we have all the time in the world for this book, let's take the mystery out of electronic invoicing, to give you a clear idea on possible options.

Notice my choice of words, "possible options". As much as I think I know all about E-Invoicing, I do not know it all (please don't let that get out). We are going to be analyzing some options... not all. However we will be discussing the most popular options on the electronic side. These options will accommodate for over 90% of uses in the marketplace.

To better understand the different types of electronic invoices we need to divide them into two camps. Camp one is called "Executed". Camp two is called "Transmitted".

Executed electronic invoices are sent by the vendor through a web portal. The vendor will have a user name and password to a website. That vendor will take the invoice that was created from the vendor's accounting system and data enter the information into a form. Once the vendor has entered the information he/she will submit the form into the portal. The software that is attached to the portal can use one of three options to distribute the information. (1) The information is sent directly into the client's approval/workflow software, document management, FTP site or the accounting system. (2) The information is collected into a standard form that can be imported to client's approval/workflow software, document management, FTP site, or the accounting system (Note the difference between 1 and 2 is (1) is a direct feed – "real time" and (2) is a download from the invoicing portal and an upload into the client's system of choice. (3) Some information is sent directly into the client's approval/workflow software, document management, FTP site or the accounting system and other information is uploaded into the client's approval/workflow software, document management, FTP site or the accounting system.

Here is the upside to an executed electronic invoice. The accuracy rate is very high because a person (not machine or software) is entering the data... this is a very good thing. I get the question all the time... "What happens if the person that is entering the information is wrong?" The simple answer to this is fire that person. Oh, sorry; Corporate America doesn't like the word; "fire". Here is a better way of putting it; Reassign the person. The other answer is to validate the accuracy of the data before the data is entered into the accounting system. This practice of validation is called the approval process. There is also a great need to control the data that is entered. Through the web portal you have the ability (if creating your own portal) to create (as required) certain fields such as, invoice number, dollar amount, ship to location, invoice date, due date, and purchase order number (just to name a few). If you are not creating your

> *This practice of validation is called the approval process.*

8 PITFALLS OF ACCOUNTS PAYABLE AUTOMATION

own portal and instead are purchasing a service that creates the portal (which is what I suggest you do), you should either buy the service that has the fields you need or buy a service that is flexible enough to add fields and make certain fields required. So to recap, the pros for the vendor executing the electronic invoice are accuracy and data control.

Here is the downside. The vendor has to do double entry. The first entry is into the accounting system/ERP system as a receivable. Most systems are constructed to produce a paper invoice, the invoice is mailed to the client and that's the end of it. If the client requires (asks) the vendor to enter the invoice into a web portal for processing, chances are you are going to get some resistance. Here are a few ideas I have used to help increase vendor adoption with clients that work very well. One is offer the vendor attractive terms. I had a client who told his vendors that if they mail him an invoice it will be paid Net-30, but if the vendor enters the data into the web portal it will be paid Net-15. It worked great! The client was able to guarantee such terms because the invoice was entered directly into their workflow/approval system, approved and electronically transferred into the accounting system within 2-3 days. Another idea is to do the entry into the portal yourself. This is the least disruptive process; because the vendors will send you the invoice as usual (a huge benefit in this process is the centralization of where all invoices are sent). Another plus to this is it will allow you to train and track the entries for speed and quality. The key to this process is to ensure that if you are not feeding the data directly into your accounting system, the workflow software, document management system or FTP site has an electronic transmission to the accounting system... you do not want to be stuck with double entry (once in the workflow/approval system and again in the accounting system). The third option is similar to the second, except the entry is outsourced to a third party company instead of having internal employees do the work. The third party company can be state side or off shore. I have had experience with both, and both have practical applications. I will tell you, based on my experience, that I have had more trouble with quality in the off shore model than with the stateside model. The outsourced option is very cost-effective; however the control of training and quality is not as high as when you have the people under your roof. Keep in mind that the "ability" level required to do this type of data entry is not the level of an accounting person. There is no accounting skill necessary to do this type of entry, unlike an AP clerk who may need understanding of the General Ledger or other specialized skills.

When discussing all these options with clients, I'm asked which one of the methods is best; one, two, or three. My answer is always, "Yes". In addition to keeping them alert, the answer proves the point that each application has its place within certain types of companies. To develop an understanding of what option may work best for your company you must look at monthly invoice volume. If your invoice volume is under 400 invoices a month and you are in smaller markets, the option of having the vendors do the entry is possible. A key to having your vendors do the entry is good communication and/or a good relationship with your vendors. If your invoice volume is between 400 and 4000 a month, option two in which you enter the invoices into an approval system is a good option. I have clients with entry personnel that can enter 400 invoices a day on top of other duties. Lastly, if your company manages over 4000 invoices a month, outsourcing the entry function is a very good option. The reason for this is that with higher volumes the entry process is a full time job. When outsourcing, you can construct agreements with outsourcing vendors to bill based on the number of invoices processed and the accuracy rate for those invoices.

To reiterate, one downside of executed electronic invoices is double entry. Although there are ways to manage this process, your organization is not eliminating steps because the entry is done earlier and sometimes by different people that may or may not be employees of your company.

Now to analyze the second camp of electronic invoices; transmitted. Transmitted invoices are by far the sexiest and most talked about type of electronic invoices. They are also the most expensive and hardest to set up, control, and work properly. We are going to look at three types of transmitted electronic invoices; OCR, EDI, and XML.

OCR, this is the granddaddy of electronic transfers. OCR stands for Optical Character Recognition. It refers to a machine (computer) having the ability to recognize letters (words) and symbols and then translate those letters and symbols into data. Simply put, it is the process of converting what is written on paper into data. The very first OCR patent was granted in 1929 in Germany and in 1933 in the United States. OCR has been around for a long time. However, the first (practical) uses of OCR were in the early 1950's with military applications. In the mid to late 1950's and 1960's companies like Reader's Digest and the US Post Office started to use the technology.

Today OCR is used in the Accounts Payable world by scanning invoices and having an OCR reader (which is embedded in software) process data on the paper into an electronic from. The data is transmitted into workflow/approval software, document management systems, FTP sites, or the accounting system. The difference and the reason OCR is considered an electronic transmission is that a human does nothing in the sending of the data. One way OCR readers work (with an invoice) is by breaking the paper invoice up into 6 zones (as an example – it does not have to be 6 it can be more or less). The OCR software will look in zone 1 for billing information and zone 2 for invoice information. (See Fig. 14) Even with the strict definition of zones, indexing of symbols and relative statements, OCR's downside has been well documented. In survey results sent out from the fine folks at IOMA (www.ioma.com) they had this to say about OCR:

> *We (IOMA) are aware that this technology, generally referred to as OCR (optical character recognition), has had somewhat of a troubled history. Some time ago, many people tried it and had bad experiences with it, especially in its accuracy rate.*
>
> *But recently, we found out that users today are happy with the accuracy rate. Almost three-quarters of users we surveyed said that recognition accuracy met or exceeded their expectations. At our Chicago conference, users were reporting accuracy rates of over 90%.*
>
> *Vendor claims: Vendors will quote accuracy rates approaching 97% or 98%, but those numbers are virtually meaningless. It's like anything you do in a laboratory under controlled conditions; the accuracy rate is totally different from when you're doing it in a real world environment. Published numbers have*

nothing to do with the appropriateness of the solution for your particular situation.

The only useful answer to this question is, "it depends on the kind of invoices you're scanning." The best test is to run images of your invoices through data capture applications from various vendors to determine which solution produces the best results. In some cases, it may be worth paying the vendors to install their systems for proof-of-concept testing at your location to obtain a real comparison of accuracy rates.

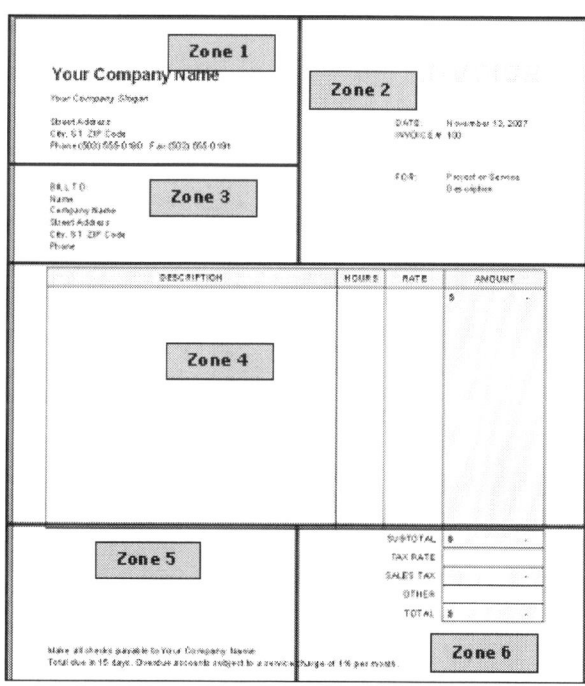

Figure 14

The email from IOMA points out the problem with OCR's accuracy in recognizing information. This problem is compounded when the information the OCR reader is trying to convert is in handwritten form. So, when considering OCR as a possible solution, ask around in AP to find out how many handwritten invoices you receive.

If OCR is the granddaddy of electronic transfer, then EDI is the younger rich charismatic uncle (kind of like my Uncle Buddy with the Betamax). EDI, which stands for Electronic Data Interchange, is a set of standards for how information is transferred from one computer system to another, and is designed to be used independent of software. I know! Computer to computer talking... it's like HAL in A Space Odyssey 2001. Well, not really. HAL took over the spaceship while trying to become human. EDI will not do that. Even though EDI is a standard transmission from one computer to another, there are many forms that can be transmitted; such as Purchase Orders, Inventory, Financial Reports and Invoices. The very first EDI standard was created in the US in 1984. It had grown to over 100,000 relationships in 1999 (Forrest Research). However, the roots of EDI go all the way back to 1948 when a man named, Edward A. Guilbert created a standard transmission of information during the Berlin Airlift[5]. The unique thing about EDI is that only the largest of businesses have been able to leverage this type of technology. EDI is generally used in manufacturing companies that have 30,000 plus invoices per month. The reason EDI is the vehicle of the elite is cost of implementation. I read of one instance in which the EDI project cost over $1 million dollars, including consulting fees, but saved the company $9 million dollars in process

---

[5] http://www.highbeam.com/doc/1G1-6336829.html

cost. That's a heck of an ROI (Return of Investment)! However, most companies do not operate under an economy of such scales. Given its place in history, EDI is the method that started the elimination of paper. It wasn't until the birth of the Internet that concepts like EDI were made practical for mid-size and smaller companies.

Lastly, in the family of electronic transferred information there is the cool young 30-something driving the hot new sports car... XML (Extensible Markup Language). XML is similar to EDI in the respect that the two are standard forms of transmitting data. However, XML became more useful than EDI due to its relatively low cost and greater flexibility. Without going too far into the technical aspects of the process, XML creates tags on each side of the data enabling the system that receiving the document to determine where to save the information in its database. Here is an example of an XML tag: <INVNUM>12345</INVNUM>. The system that produced the information put the 12345 between the tags so the system receiving the information will know that 12345 is the invoice number. There could possibly be dozens and dozens of tags on a single XML file transmitting thousands of pieces of information from one system to another. An additional advantage to XML is the ability to take documents (such as a CSV file or tab delimited text file) and convert them to XML.

So what's the down side? You must have a mechanism to send and receive XML information. This requires that you employ a service and/or person to create the files and tags within files that can be read by the receiving party. To keep this in proper perspective, the cost to employ the XML service and/or person is much less than the cost to employ an EDI service or person.

We have talked about two camps of electronic invoices, Executed and Transmitted. To be clear, Executed electronic invoice requires that a person is setting up or doing data extraction into a piece of software whereas Transmitted is a machine to machine communication of data. Here is the big question: What is the best option for an automated AP process? Well, as you can imagine, the answer is not as simple as my making a judgment call on one option over the other. I have worked with each option successfully. To get a better understating of what might work for your company, I have devised a list of questions to answer. Your answers will help you to determine which option best suits your needs: electronic form vs. electronic data.

Question:

Section 1 - Corporate Matrix:

1. How many invoices does your organization manage per month? _____
2. How many full time (note: add part timers together) AP staff do you employ? _____
3. Divide line 1 by line 2 to get the AP staff per invoice per month: _____

Each yes is worth 25 points: Total _____

Section 2 - Technology:

1. Are there any automated processes running in your accounting system? ____
2. Is there internal IT staff within your company? ____
3. Are there programming resources within your company? ____

8 PITFALLS OF ACCOUNTS PAYABLE AUTOMATION

First Edition 2011

4. Is your accounting system customized to your needs? ____

Each yes is worth 25 points: Total _____

Section 3 - Logistics:

1. Do your invoices arrive to a single location from the vendor? _____
2. Is there staff dedicated to processing the invoices? _____
3. Are you receiving invoices from vendors in an electronic from? _____
4. Are you receiving invoice from vendors through electronic data? _____

Each yes is worth 25 points: Total _____

Section 4 - Current State:

1. Are invoices ever lost? _____
2. Are invoices ever paid late? _____
3. Do vendors call more than one person for an answer? _____
4. Are invoices ever paid twice? _____

Each yes is worth 25 points: Total _____

Totals:

Add section 2 and section 3 _____

Subtract the total from section 4 from the above total _____

Divide section 1 with the above total: _____

Recommendations:

50 +	First – write your own book! Then leverage internal development to create an XML transfer with the vendors with whom you have a high volume of invoices. Consider using your side of the XML relationship to capture and convert EDI transactions on both the invoice and payment side of AP. For those invoices that are not transmitted with data, develop a processing center to scan and OCR the data that can be read by your XML receiver. Keep in mind that your company is in a good position to leverage transmitted electronic data to keep your AP staff lean and cost-effective. Note: Outsourced electronic data from as many locations is the best option for your company.
43-49	Leverage internal development resources to create XML transactions with the top 10% of volume vendors. Consider an OCR component to your indexing process. Keep in mind that your company is in a good position to leverage transmitted electronic data to keep your AP staff lean and cost effective. Note: Some but not all outsourced electronic data with internal processing as a supplement is the best option for your company.
36-42	Consider outsourcing part of your invoice processing to a third party company that is stateside with the invoices entered into an established approval/workflow web portal that stores the data and has an electronic feed to your accounting system. Also, introduce the web portal to vendors with which you have good relationships and communication.
22-35	Check with a third party company if any of the following items are not in place: Scanning and processing invoices to a web portal, saving invoices in a common document management system, enabling an electronic approval system to track and speed up the invoice approval process, creating an electronic feed to your accounting for fully approved payables, and centralizing AP processing functions for a single point of vendor contact.
17-21	Create a plan to centralize AP functions and purchase a service that will enable invoices to process into an automated approval workflow system. Develop SLAs (Service Level Agreements) through the approval system to eliminate lost invoices, speed up late paid invoices and stop double paid invoices. Redefine roles within AP to

	create value added tasks that serve other departments or consider eliminating AP as the organization becomes more efficient with an automated approval process.
10-16	Create a plan to centralize AP functions and purchase a service that will enable invoices to process into an automated approval/workflow system. Develop SLAs (Service Level Agreements) through the approval system to eliminate lost invoices, speed up late paid invoices and stop double paid invoices. Note: Electronic forms and your company controlling the entry is your best option.
0-9	Create a plan to centralize AP functions with secure electronic storage of invoices. Note: Getting centralized with no electronic data or form is the best option for your company.

(Note: for a printable version of the above chart/questions go to http://8pitfalls.com/ws/worksheet6.pdf)

These recommendations stand on their own. However, keep in mind that there are always exceptions to the rule. The questions and the scale is only used (it's a good use) to give you an understanding of where your organization matches up against other organizations. When you know where you are on the recommendations, you are able to set more realistic goals for an automated AP process. Merely understanding the difference between an electronic form and electronic data can help you to communicate with your vendors and with internal management that you want the newest and greatest. So, if you are an organization with 500 invoices a month and 3 AP staff trying to find lost invoices, and your CEO comes to you after a board meeting and asks you to look into EDI... ask him if, in the late seventies, he owned a Betamax!

Bonus 2!

Who Really Wants To Pay Early?

Automation has a tendency to create what I like to refer to as "happy accidents". A happy accident is something profoundly helpful that the company did not consider or factor into their automation project. There are many happy accidents, but the one that directly relates to the supplier is vendor discounts. Vendor discounts are the unsung hero of the AP department. Discounts fly completely under the company's radar for several reasons. One, AP department leadership is not in the business of looking for sources of income (this, by the way, is a minor one). Two, historically, the approval time has been too long to even think about discounts. Thus the happy accident! When AP is automated, the approval time can go from 15, 30 or 45 (I have seen it as high as 95) days to (my clients average) 2.8 days. An under 3 day approval cycle makes capitalizing on a 2% net 10 not only possible but it gives you 7 extra days.

Make AP a Profit Center:

Here are some numbers to consider from a mid-size accounts payable department (note: EPD = Early Payment Discounts)

Monthly Discounts	
Monthly Invoice Transaction Volume	4,500
Average Invoice $ Amount for EPD	$1,539
Percentage of Vendors with EPD Offerings	16.80%
Average % of EPD Discount	1.28%
Total Monthly EPD Opportunities Available	$14,798
Total Annual EPD Opportunities Available	$177,586.

Figure 15

Following Figure 15's numbers, this company has 4,500 invoices per month. Each invoice averages out to be $1,539 across the 4,500 invoices per month. Not all vendors offer an early payment discount. I have found that the percentage of vendors offering a discount is less than 17% (16.80%). I have also found that the average discount per vendor is 1.28% for various times (Net 10, Net 15... but better than Net 30). Doing the simple math the total monthly discounts available are $14,798 or $177,586 per year. In the real world an AP department would not be able to capture every discount every time, but a well-tuned automated AP department would be able to capture 80% of the discounts. What would be the reaction of the leadership of your company if you told them you would be able to provide them with $150,000 in savings each year? Of course this dream only comes true once your AP process is automated.

And Finally...

Dealing with supplier flexibility is the key. Giving suppliers options to deal with you and having each of those options in your AP Automation tools creates an environment that cuts down on forced changed, which can become very expensive to your company. It is also important to be educated on proper file formats when it comes to communicating with your suppliers. But the most unsung hero of the AP department is the ability to capture suppliers' early payment discounts. After you have automated and you approach your leadership with a new savings opportunity, make sure you ask for a percentage of the return as a bonus!

Chapter 6

The Training and Going

Pitfall – A Proper Perspective On Training Will Keep You Out Of The Pit.

After giving a project presentation or a demonstration, I am often asked how I felt about it. My usual reply is that, "It's software" so it has no feelings therefore I don't either. Now that's not entirely true; I have plenty of feelings. Like right now... I feel bad about telling this next story.

Working in Mississippi on an AP Automation project I was near the end of the engagement. The last thing left was training. The client decided it was best for me to do the training, which was fine with me (later you will read that having the software provider do the training may not be the best way to do training). There was a good size room; tables and chairs lined up classroom style, and a projector and screen at the front. Every seat was filled. The training was scheduled to have a demonstration first and then small group sessions afterwards. During the large group session, a well-meaning woman in the front of the room kept asking if it was possible to move a button to another section and then change the color to a different one. This went on the entire training. It got to the point that I needed her to learn the software before making suggestions on how to improve it. So I said, "What we're working on is the ability for you to stare at the computer screen and have it do exactly what you want it to do; but until that day comes we need you to learn how the software works first." The entire class erupted with laughter and I didn't hear from my new friend again. I know you might be thinking if I said that to you, I would have lost a client. Luckily, it was well timed and the client actually saw it as being very helpful, because training on new software is difficult but it is even more difficult when that software creates new processes. In this chapter, we are going to outline how to train both software and process change successfully, without putting the entire automation project in jeopardy.

The interesting thing about AP Automation is its uniqueness. Because it is new technology, there is no benchmark set for implementation. AP Automation is unlike an accounting system, in that an employee would say, "Here we go again. At my last job we changed accounting systems and it was a nightmare," and be able to list the reasons why it was a nightmare; even giving specific areas for the project management team to watch out for or be aware. Early in my career the notion of using the internet to transmit and approve invoices was so foreign I mostly handled questions about the security of the information. More recently the questions have been around process improvement and becoming more efficient or streamlined. In the past, it was also difficult to benchmark a solution because competition was scarce. 2007 saw an evolution of AP Automation that introduced more knowledgeable consumers as well as competition. However, it still remains that AP Automation is unique to an organization. For this chapter it will be mostly assumed that shuffling paper is the current method and the training we are going to be discussing is the change from paper to an automated environment.

> The interesting thing about AP Automation is its uniqueness. Because it is new technology there is no benchmark set for implementation.

Out With the Old

Leaving behind an old method and entering into a new territory involves change management; which conveniently is one of those terms that is thrown around at each meeting for multiple reasons. The dictionary defines change management as, "Techniques that aid in evolution, composition and policy management of the design and implementation of an object or system[6]." I believe the key word here is evolution. The downside to evolution is (typically) growth, and, with growth, growing pain. Organizations accept change differently. If your organization scores below a four in the "change factor" questions in Chapter 2, change may be very difficult. When connecting the dots between change and training, it is naive to think that if you just train on the software then the change part will take care of itself. Meaning, if during the training phase of your process you teach people how to point and click then everything should be great. This is never the case.

There are two types of change management; software and process change. This chapter will define both software and process change as well as guide you on successful training techniques.

Plan for the Change!

I have found there is a lot of fear around software prior to training. Questions like, "How will I code my invoices?" Or "Where are the filing cabinets? Or "What's so bad about paper?" If you look at the nature of the questions, it's not fear of the software as much as it's fear of the unknown. One of the best places I have found to control fear and help the people in your organization understand what life is going to be like is with the use of "Super Users". Super Users are well trained and respected employees that get to know the software and process before the software rolls out. In Chapter 3 we discussed the creation of a testing group that would eventually become administrators or

---

[6] http://dictionary.reference.com/browse/change+management

Super Users. Super Users are pivotal to the training process. They are people on the inside when the discussion of new software comes up to say, "It's not as bad as you may think." Or "I felt like that in the beginning but once I approved a few invoices the fear went away." That's why it is important that a Super User be the person who embraces technology or is not scared of software's unknown factor. The next thing to hedge the fear before training is to pass around a FAQ sheet. The sheet below has been provided to serve as an example of the list of questions and answers to circulate after the original introduction of change to your user group.

General Problems
• Who do I contact if I cannot login or forgot my password?
• What is a system administrator?
• What is the difference between a system administrator and a portal administrator?
• Am I able to use the system at home or away from the office?
• How do I use Help and what type of information can I expect to receive from Help?
• How do I set a subscription notification?
• Where are all the invoices stored?
• Will I need to keep a copy of the invoice at my desk?
Account Codes
• How do I split accounting codes?
• Can an invoice be divided between operating codes?
• Will the system allow me to split the invoice by percentage or dollar amount?
• What happens if I don't know my accounting codes?
• Am I able to code the actual invoice?
• Who has the permission to code an invoice?
• As a manager can I allow someone in my company to code all invoices?
• Is there a way to ensure that an invoice is fully coded before going into the accounting system?
• How do I add a code?
• How are codes entered into my accounting system?

Notice the motive of the FAQ is to prepare the end users with the basic questions and answers. This will set the expectation on how this new software is going to be used. You will find a complete FAQ at the end of this chapter for your use.

More Planning for Change

Process change training is different from software training. The difference comes from nothing being tangible that the end user can do before the project goes live. Previously, the invoices going to the department's administrator were no longer going to show up. Instead all invoices are to be directed to a processing center's P.O. Box or a vendor provided file and then directly to the software. Just like software training, process training can be introduced by the internal super user as well as a FAQ. See an example below:

FAQ's For Basic Users

Q: Why is <company> changing where invoices are sent and how they are processed?

A: <company> is changing its approach to invoice processing in order to improve the efficiency of the payment approval process. We anticipate that having one centralized location will lead to a decrease in lost invoices, which delay payments to our vendors. In addition, this new process provides an expedited invoice review, which could facilitate quicker payments to our vendors.

Q: How does this change relate to the <accounting system>?

A: The improved invoice payment process is in addition to <accounting system>. It is meant to enhance the entire AP work flow process and works in conjunction with <accounting system>.

Q: How will our employees be trained to the changes taking place?

A: Our employees will be trained at a formal meeting on <date>.

Q: When will these changes take place?

A: The process change will be effective as of <weeks before Go Live date> for the vendors providing goods and services to <company>. Nevertheless, these vendors may commence sending invoices to the new remittance address immediately. We will commence scanning and processing invoices to our employees attention and you will then review and approve these invoices online, for <company> on our new system as of <Go Live date> coinciding with the receipt of invoices from the vendors to a centralized location. Failure to do so will impact both the speed and cost of the process as well as the financial results of actual versus budgeted expenditures for your buildings.

Q: How long will these changes take to implement for the vendor?

A: If their invoices are normally mailed, a simple change to the mailing address is all that is necessary. If the vendor's invoices are dropped off at the office, then the vendor is required to mail their invoice the new AP remittance address. Vendors will be receiving communication in advance regarding this change and will continue to do so until they are in compliance.

Q: How are vendors' invoices to be sent to <company>?

A: Invoices should be remitted using regular First Class mail. If a vendor needs to use overnight service method for delivery, they must use the U.S. Postal Service's Priority Mail. FedEx or any other overnight delivery service will not be accepted at a P.O. Box address, and invoices must be received at our P.O. Box.

Q: What new information should vendors include on the invoice to expedite payment?

A: A majority of the invoices already contain the required information. However, all invoices must include the NAME of the <department/location> where goods and services were provided. The address should be included in the invoice if applicable. If the invoice is a result of a service request, work order, or purchase order, the

> reference number must be included; otherwise there are no other changes.
>
> **Q**: What happens if a vendor forgets to make the change?
>
> **A**: Failure to comply with these changes could result in a delay of the vendor's payment and impact the financial results on your actual versus budgeted expenses for your department. We will continue sending our letters to the vendor regarding the change and will copy you on this correspondence so that you may follow up with them.
>
> **Q**: Who do I contact with questions?
>
> **A**: We will inform you of a specific e-mail address and AP processing phone number for contact, once we have established our remittance address.

Lastly, I have had success with small/large group training, outsourced training, individual training, phone training (with or without web) and email training. Here is a definition of each type of training.

- Large Group Training – When people think of training the large group setting is what first comes to mind. The large group is more than 10 but can be as large as hundreds listening to an over-view of the software. This type of section is excellent for change management where the purpose is explained and people can benefit from group questions and answers. The large group setting is not good for specific software training, especially where people are in front of computer terminals. The slower people will hold the group back from a comfortable progression.
- Small Group Training – Usually 5-10 people. Small group training is good for change management, but

is better for specific software training in front of terminals. To do correctly, you should coordinate people of like jobs and abilities and tailor the training to the group. The small group will allow the instructor to look over shoulders and give specific instruction to individuals without slowing down the progress of the entire group.

- Outsourced Training – Just as the title implies, this is your company contracting with a company that does training as their business. The upside to using contract trainers is – training is their business. You will be able to leverage their training expertise. The outsourced training company may have facilities that are optimized for a positive training experience. The downside is the trainer may not know your company's culture well enough to pull off the change management piece of the automation transition.
- Individual Training – This is one-on-one training. This is usually done at the person's desk with the trainer looking over the shoulder of the person being trained. This type of training works great for smaller companies, and those companies that leverage the approach of the "Super User" written about in this chapter. The down side to individual training is (if you let it) it can take a long time and involve a lot of travel which will make this approach expensive.
- Phone Training – Believe it or not, this is an effective use of training time. As with individual training, this is personal one-on-one training with the trainer walking the trainee through the leaning process. I have done this. I am watching the person on my computer and I have also done this

- while I am giving verbal queues to the person on the other line. The verbal queues require a person skilled in the art of describing detailed steps. The outcome is the trainee physically has completed all tasks associated with the new procedure when the phone call is completed.
- Email Training – Under the right circumstances, sending out instruction by way of email can work. When the software and the new process are not complicated, the people being trained are savvy and not resistant to change, and the software has been tested well and the process is stable, this is a viable method.

It has been my experience that the most efficient and cost effective way to train is through smaller webinars (www.gotomeeting.com or www.webex.com). Also, the training is best done by one of the Super Users/Administrators rather than a trainer from the AP Automation provider. You might think that the best option is to let the automation provider do the training. The flaw in this thinking is that training is done from the software perspective, not from the process perspective. The reason for this is the process perspective deals with business change, which is not received very well when a person outside the company is presenting it. It has been my experience that of the overall time, 20% is spent on software training (the physical points, clicks and outcomes of the software) and 80% on process change training. Understanding this concept will be very helpful in your organization's ability to spend the correct amount of time preparing and executing a useful training plan. Also keep

> *If AP Automation is trained in a spirit of "Let's give this a try," it will create the seed of rebellion.*

in mind that giving end users/employees the option, their natural tendency will be to gravitate towards what they are familiar with. I am not a big fan of mandatory change, however if AP Automation is trained in a spirit of "Let's give this a try," it will create the seed of rebellion. Rebellion is the number one cause of a process change crashing and burning. It is critical to have a firm commitment from leadership within your organization to ensure there is no rebellion.

Re-Training

Don't overdo it. It's important to keep the automation piece of accounts payable within the proper perspective. AP Automation training is not like accounting system training. AP will not reach the number of people and the number of tasks needed to be done within the company like an ERP system or a standalone accounting system will.

There is something to be said about training once on the basics and then having follow-up training later on the process. Training a person once on a new process is difficult from the end users perspective because they do not know what they do not know; therefore they also do not know the correct questions to ask. However, after a certain period of time, say 30 to 60 days, the end users will have questions and even possible improvements to the process. It's a great idea to create an atmosphere where the end users feel free to give feedback and have that feedback seriously considered.

More Feelings

I started this chapter with feelings and I'm going to end it that way. This time I do not feel bad, but somewhat conflicted. Over the years of kicking off projects, training and rubbing elbows with end users, it amazes me how often the conversations about the software or process change are emotional in nature. Contrast that with a technology-driven person or group void of any possible emotions (somewhat kidding about the void thing). The dichotomy of the emotional and the emotionless can create the perfect storm. Change can be difficult for some and implementing software may be black and white to others. My suggestion is not to mix the two in training. Choose your trainer(s) or Super Users wisely by considering who they are and who they're teaching. Once you have the right person or team, make concerted effort to separate software and process change, each having their own goals and outcomes. Keeping the training in the proper perspective, knowing your audience and creating a plan tailored to that audience will keep you out of the pit!

Full FAQ

As promised, here is the full version of the FAQs for end users.

General Problems
• Who do I contact if I cannot login or forgot my password?
• What is a system administrator?
• What is the difference between a system administrator and a portal administrator?
• Am I able to use the system at home or away from the office?
• How do I use Help and what type of information can I expect to receive from Help?
• How do I set a subscription notification?
Account Codes
• How do I split accounting codes?
• Can an invoice be divided between property codes?
• Will the system allow me to split the invoice by percentage or dollar amount?
• What happens if I don't know my accounting codes?
• Am I able to code the actual invoice?

- Who has the permission to code the invoice?
- As a manager can I allow someone in my company to code all invoices?
- Is there a way to ensure that an invoice is fully coded before going into the accounting system?
- How do I add a code?
- What codes does the system accommodate?
- What type of codes can I put in the system for my users to access?
- Can I associate certain codes to entities?
- How are codes entered into my accounting system?

Allocation Codes
- How do I set up an allocation code?
- How do you edit an allocation code?
- Can I limit who sees particular allocation codes?
- What is an allocation code?
- Can an allocation code be set up for codes other than properties?
- What is the maximum number of allocations the system will allow?
- Who has the authority to assign an allocation?

Importing Data
- How do I get the vendor import format?
- How can I add a GL code into the system?
- How do I add a user into the system?
- What type of data can I import into the system?
- Is there any special format or file that I need to import data?
- Who has the ability to import data and can that permission be given to anyone?
- Can I update the data that is currently in the system?
- What happens when I attempt to import information that is already in the system?
- How do I import properties and property codes?
- If I import incorrect data can I do a mass delete?

Invoices
- How do I change the amount on an invoice?
- How do I change the vendor on an invoice?
- How do I change the image on an invoice?
- What happens when I dispute an invoice?
- How do I add a comment to an invoice?
- Who is able to view invoices?
- After the invoice is approved, can I go back and view the invoice?
- Why is there duplicated information above the scanned invoice?
- What are the meanings of the tabs: image, vendor, line items, workflow, history and comments?
- Am I able to restrict information on the invoice?
- Can I delete an invoice?
- Why is there a save button at the top of the page and a save button on the workflow tab?

8 PITFALLS OF ACCOUNTS PAYABLE AUTOMATION

- Who do I contact if I am unable to view an invoice image?

Pending Approval Queue
- How do I sort my pending approval cue?
- Can I approve more than one invoice at a time?
- What is the Pending Approval Queue?
- What do I do if the Pending Approval Queue does not show up?
- Am I able to view invoices from other property managers in my Pending Approval Queue?
- Can I view other user's Approval Queues in my organization?
- How do I view codes from the Approval Queue?

Users
- How do I change my password?
- How do I reset a password?
- How do I set up my email subscription?
- How do I assign users to properties?
- How do I assign users to roles?
- Can a user be assigned to more than one role?

Workflows
- What are the workflow rules an invoice follows when entering the system?
- How do I enforce full coding on invoices?
- How do I change an existing workflow?
- How do I create a new workflow?
- How do I set up conditional workflows?
- How do I assign a workflow to an entity?

(Note: for a printable version of the above chart/questions go to http://8pitfalls.com/ws/worksheet7.pdf)

Chapter 7

The Evaluation

Pitfall – AP Automation Is Not A One-Time Project. In Order To Stay Out Of The Pit You Need To Be Continually Improving.

Roller Coaster

Some people love roller coasters. They get a big kick out of them. Me not so much, however I ride a roller coaster every time an AP Automation project goes live. With zero exceptions it starts in the training phase with fear, then acceptance, then joy of the possibilities, then anger with the change, then comfort with that same change, then desire to improve missing pieces, and finally it becomes part of life. (Fig. 16)

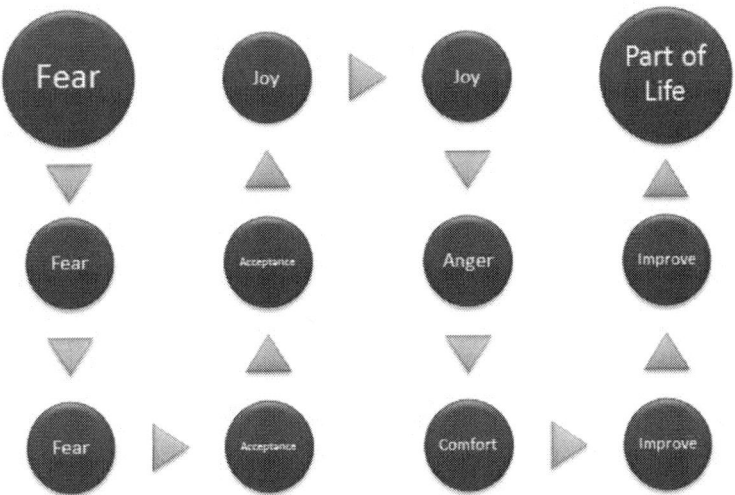

Figure 16

Even though the steps are the always same every time, the timeframe it takes to get from fear to part of life is different. Some companies may go from fear to part of life in two weeks; others might take 6-9 months. Once out of the woods and into the part of life phase the process, AP Automation supplier, and goals of ROI need to be evaluated. This chapter will give you the steps and suggestions on how your automation project is going and where you can make it better.

Take a breath!

All the scoping, building, testing, configuring, and training is done... take a breath in and then out, and now it is time to get back to work. One of the problems with an AP Automation project is too much time, effort and attention is focused on the Go Live date. All the preparation, all of the overtime, all of the anticipation and the day of Go Live comes and goes, just like that. Some companies have a party, some high fives, others a big "I am glad that is over" sigh. However; the work is not over. At the point of Go Live, the work has just begun. I have been on projects where the client has put a check mark in the done column, walked away from AP Automation and was surprised six months later when the process was in shambles. This is the worst case situation. "Not as good as it could be" is a more accurate situation. Optimizing is extremely tricky, because good is good but great is better than good.

> At the point of Go Live, the work has just begun.

What it takes to get from bad to good to great is effort. Illustrated in Fig. 17, the effort that it takes to get from bad to good is much steeper than going from good to great.

Figure 17

To help your company get from good to great, I have two suggestions. First, survey your users. Second, have the system tell you where to improve and how you are doing. (The second one is quite revolutionary.)

Here Is A Survey To Fill Out:

I was an original user of Salesforce.com. It was my introduction to SaaS (Software as a Service) and I loved it. What I loved more than the software was the surveys they would send out in the early days. I know this sounds strange but I like filling out surveys. I enjoyed the free stuff that came along with filling out the different surveys. I'm not suggesting you give out free stuff (although it will help). I am suggesting, however, to ask your users in a non-threatening anonymous way how the process can be improved. Noticed I typed "process". You can ask for software improvement but the responses may be numerous and off the chart (moving buttons, changing colors, rearranging the screens). Your survey needs to have a disclaimer somewhere to let the user base know each suggestion is needed and evaluated but it will not be possible to use every suggestion. Because I mentioned process, the outline and presentation of the survey will focus on the process; if you greatly desire to survey the software feel free to add that information.

Types of Question

I have always used questions as tools. The answer to a question can be used in two ways. One - Look at the actual answer to the question. This is the basic approach that most people take. An example is: "How often do you login and approve invoices? Twice a week. Two - Ask a question that will allow you to gain insight on how it is answered. An example is: "Given the current roll out, would you prefer to keep going with automation or would you prefer to go back to the old way?" If going back gets more of the votes it doesn't necessarily mean that your company made a mistake in automating but it does mean that you need to get a better understanding of why the rollout was so poor. The answer to that question could lead to improvements to the process. My suggestion is to survey your people in three phases. (1) Immediately after Go Live. (2) Six months after Go Live. (3) One year after Go Live. You can go past one year if you think you're getting good data; however, the life cycle of automation spans about a year to eighteen months, and if you do not make any adjustments after Go Live, the process will not have evolved. Below are a few suggested questions along with analysis of the questions.

Phase 1 – Questions after Go Live:

On a scale of 1-5 with 5 being the best; answer the following questions:

1. How would you rate your experience with the training?
2. Does the process and software fit your needs?
3. Is the software easy to use?
4. How likely are you to recommend this software/process to a colleague?
5. Did the rollout meet your expectations?

Please answer the following questions in the sections provided:

6. Does the new process leave anything out that you need in order to do your work?

7. Will this process improve the way you do your work? Please explain.
8. If you had the rollout to do over again what would you change?

Phase 1 – Questions after Go Live – Analysis:

It is best to use 1-5 as the scale. This is to eliminate what I call the "7" factor if you use a 1-10 scale. The "7" factor is a no man's land of information. On a scale of 1-10 people naturally gravitate towards 7 because it is non-controversial. In this survey it is important to get useful information. A 1-5 scale will provide clearer results for the type of questioning you will be conducting. Each level is explained the same using: Great job, Be Aware, Neutral, Bad News and Panic! The words after the initial designation will explain the analysis.

Question 1 - How would you rate your experience with the training? (Below are averages – total up all responses and see where you fit on the scale)

Number	Action
5	**Great job** - You did well on the training
4	**Be Aware** - that there was a slight dislike with the trainer or the material – don't panic
3	**Neutral** - not the best place to be because it does not give you enough to re-train and it does not give you enough to seriously consider the process is broken, what it does mean is you need to ask more questions to find out if it is the training, process or software.
2	**Bad News** – find out if your people hated the trainer and their materials or if it was the subject matter. In either case – more training is needed.
1	**Panic!** - This would indicate that the scale got flipped and everyone thought "1" was the best (which creates another problem) or you are way off the mark. You need to look at the overall project and consider starting over.

Question 2 – Does the process and software fit your needs?

Number	Action
5	**Great job** – the vendor section was done well and the scoping of the project as well as the change management was done well.
4	**Be Aware** - there may be a few things missing but don't panic.
3	**Neutral** - Be careful about this one, it might mean they don't view their needs as important. If that is not the case separate the question to isolate where you need to strengthen the process or the software. The answer to later questions can help you in this decision.
2	**Bad News** - As above in the neutral section separate the process from the software. Then devise a plan to analyze your current situation so you can re-scope the project. This doesn't mean you will have to do a full blown re-implementation but chances are you will have to do some type of re-implementation.
1	**Panic!** – This is harsh, but start all over.

Question 3 – Is the software easy to use?

Number	Action
5	**Great Job** – The selection or build was done well.
4	**Be Aware** – There may be a few gaps but none that are show stoppers so, don't panic.
3	**Neutral** – The division here is if you build the software you may need to change or add a few things, if you bought the software you need to make the suggestion to the vendor for the changes or additions.
2	**Bad News** – The same split as above, however if you built the software you need to consider bringing in someone with more experience to do the updates or modification. If you bought the software, alter the vendor that you have serious problems.
1	**Panic!** – Continuing on the split, if you built it go out to market and find a vendor who can replace the software you built, and if you did buy the software put them on notice that you are switching vendors.

Question 4 – How likely are you to recommend this software/process to a colleague? This is one of those indirect questions. If you were a software company, you would be looking for leads. Assuming you are not, you are looking at an overall satisfaction level, meaning: The world should know about this (5)! I wouldn't do this to my worst enemy (1).

Number	Action
5	**Great Job** – Their overall satisfaction is good
4	**Be Aware** – There may be a few things lacking but overall they are happy with the outcome.
3	**Neutral** – Your analysis should begin using the section of this book for find where their dissatisfaction comes from, the planning, vendor selection, scope of the project or the training.
2	**Bad News** – As above find out in which phase you went wrong and create a plan to change. A 2 is a red light, be proactive about changing the red to green.
1	**Panic!** – Something has gone horribly wrong; this will not be a surprise because the rest of the questions on the survey will be very low. However, if the rest of the scores are not low then you may have a company who does not have colleagues.

Question 5 – Did the rollout meet your expectations? The answer to this question will help focus on change management.

Number	Action
5	**Great Job** – The project was scoped well and the training around change was done appropriately.
4	**Be Aware** – No need to panic getting a 4, you know your company; they may be hard to please. If not, look into a few additional help files, blogs or FAQs.
3	**Neutral** – As in question 4, find where the change breakdown occurred. You will need to rollout additional information, emails, or blogs. Help files or FAQs.
2	**Bad News** – As in 4, be aware you may have an environment that is not used to change (refer to Chapter 2 and the factors of change to help). Getting a 2 means you need to focus on more proactive measures of change, that

	could mean webinars on the change or face to face meetings and large group meetings with super user testimonies will help.
1	**Panic!** – Don't give up, but organize yourself. Take the responsible approach that there may be something you have not explained well enough and start over with face to face meetings in addition to printed materials.

Question 6 - Does the new process leave anything out that you need to do your business?

Analysis: No hidden agenda here, this question is to insure you did not missing something. Look for things like a workflow exception, a type of invoice that needs special handling, or a situation in which an invoice needs back up that is not accommodated in the workflow. If this comes from enough people or was a blaring mistake, add it to the process and let the user know of the change.

Question 7 - Will this process improve the way you do your business? Please explain

Analysis: Clearly the "please explain" part of this question is what you are looking for. This can give insight as to whether the conception, execution and outcome of the project vision meet reality.

Question 8 - If you had the rollout to do over again what would you change?

(Note: for a printable version of the above chart/questions go to http://8pitfalls.com/ws/worksheet8.pdf)

Analysis: The answers to this question will give insight on actions moving forward. Answers like, "I would not have done it" or "I would forget it ever happened" will be handled with the scale of 1-5 answers, meaning your scores would be low. Answers that are helpful in this section are: "I would train the managers and assistants separately", "I would not have done the go live so close to the quarter end" or, "I would have cleaned up the vendor file or GL before doing AP".

I am not going to outline questions for the six and twelve month evaluations. You can craft that yourself with the principles I outlined at the beginning of the section. I will suggest, the six month evaluation questions should be similar to the set of questions outlined above only in the format of six months later. This will allow an opening for the after one year question to focus on improving the process and change management.

Lastly, there are many outlines designed to help you send out the surveys. I highly recommend choosing one that uses the internet and even sends out email reminders. This way you can get the survey out quickly, with ease and, depending on which company you chose to go with, the software may have calculations and analytical tools to help read the answers. Besides, how awkward would it be if you send out a paper survey to see how your paperless automation system is doing?

What would HAL do?

I love science fiction movies. In 2001 a Space Odyssey the computer (HAL) takes over the ship and thus starts the battle against man and machine. Don't be afraid, because this section of the chapter is going to explain how the software and process you have put in place will tell you exactly how you're doing. Before you freak out, your software will not start talking with you (at least not yet) and when it does this chapter will be updated, but until then this section is going to help you look for clues to improving your process and software within the actual process and software. First, the reporting engine you use will need to have the ability to report on all aspects listed below. If that's not the case, I suggest you change your automation vendor or invest in a business intelligence solution. For improvement it is critical to have a reporting engine that will allow you to see user, vendor, workflow and process activity (just to name a few).

Over time (6-12 months) create and run the following reports:

- How many times were the workflows reassigned?
    - Which ones?
    - By whom?
- How many times were the GL codes changed from either the default or in a later step of the approval process?
- Benchmark at the beginning of the project the percentage or run your re-class report to see the change over time.
- How many times has someone changed a password within a given month?
- By role or job title run a cycle time (how long it takes to approve invoices) report.
- How many times was the help file accessed by which job title, or workflow step?
- How many times was the software down or unavailable and classify as an outage?
    - Software Problem?
    - Service Provider – Internet?

- Service Provider – Citrix or WAN?
- Internal problems?
- Network problems?

Once the data has been collected it is time to analyze what it says (yes data can speak). With these reports, you've gathered process, software and maintenance data. From the process side, finding out what type and how many times a workflow was reassigned will let you know if you need to build a new one. Here is an example:

In a capital expenditure workflow there are 200 invoices a month. Out of those 200 invoices 50 are reassigned to an additional approver that is not in the workflow, meaning that 25% of the time the workflow is lacking. This would indicate that you may want to create a new or update the current workflow to include the additional approver.

Another example is you have a workflow that over 6 months' time has never been used or only has 2 or 3 invoices within that timeframe. This indicates that the workflow may be useless and needs to be evaluated further and perhaps de-activated.

An example of a software improvement involves accessing help files. Let's say your report finds the help files were accessed by a certain job title at a certain point. It's important to recognize that more than one person in the job title is doing the same process at the same time in the workflow. This would indicate what the software needs and the person/job titles ability or competency is not matching up; which means the software construction is probably lacking.

Lastly, maintenance is gathering the data that revolves around up time. If your service provider is down a few times a month, then a change is needed in the provider or you will probably need to make a greater investment in your infrastructure. Unscheduled maintenance needs should be kept to once or twice a year (if not less). Anything above once or twice a year is going to cause a slowdown in the process as well as a very unhappy user experience.

I am not suggesting that the reports I have outlined are all you need to run for your company; they are a guide using the most common reports I've encountered. My goal is to give you an idea that the process itself will let you know how you're doing by running and evaluating data. This supports a statement that I mentioned in the first chapter: In a paper driven process you have no data to formulate a plan, but once you are in an electronic/automated process you have real data that will outline improvements.

> *In a paper driven process you have no data to formulate a plan, but once you are in an automated process you have real data that will outline improvements.*

Conclusion:

I am hoping that the most ground breaking part of this chapter will be that the software/process tells you how you are doing. Data is the most underutilized factual information available in automation. But, do not look past the obvious in evaluating your people. Your people will tell you what you need to know; sure, you may have to labor though a few people who hate everything (especially technology), but in the end you will have good solid information to continue to improve your AP Automation platform.

Chapter 8

The Future

Pitfall – Having An Idea Of What Is To Come Will Keep You Out Of The Pit.

Nostradamus:

When I was in High School HBO ran a program on Nostradamus. It was the story of a monk that claimed to be able to know the future. He predicted the end of the world, and even gave a day and specific time. This information quickly traveled through Watauga High School with anticipation, panic swept through the halls and people hid under desks out of fear. That was 1985, and guess what? The world didn't end...some say music did, but the world went on. I am not a Nostradamus expert, do not care to be one and really do not want to know about his predictions. I do know this about Nostradamus; he can't predict the future...no one can. Some people have better instincts than others, and correct business, economic data and information does have the ability to aide in creating good instincts, but the future is undefined. In this chapter, I'm not going to predict the future. Simply give my opinion on how to be prepared for what it may hold.

A Brief Company History:

Those who do not know history are doomed to repeat it. This is a very true statement. Emerging technology has taken a radical path because of need, ability, competition and desire (Fig. 18). Strictly from a technological perspective, need, ability, competition and desire doesn't compute. Technology loves numbers and black and white situations. Technology doesn't work well with emotions; however, the emergence and creation of technology is fueled by emotions. I was a witness to this first hand during my introduction to automated AP. Earlier in the book I had referred to "happy accidents". My introduction to AP Automation was certainly that, a happy accident. The rest of that story comes from when the President and CEO of the company I was working at the time, met with a client that said, "If you're doing the PO electronic, you need to do the invoices electronic." The CEO's next statement was along the lines of "Would you pay for that?" The client answered yes and the rest is history. I give those two gentlemen a lot of credit for capitalizing on such a prime opportunity, and as the AP side of our business started to improve and grow bigger and better based on client feedback as well as the realization that the PO side of the business was not all that it was cracked up to be; the company's revenue soon became completely based on the selling of the AP Automation with an add on PO module.

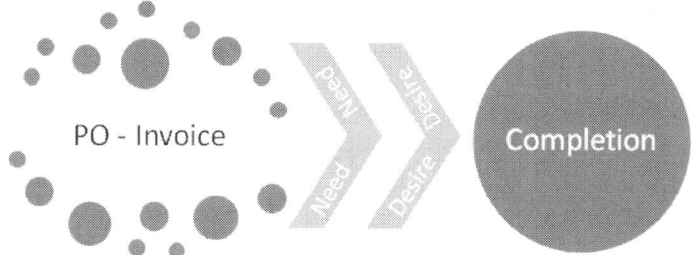

Figure 18

Note in Figure 8.1 how the progression of the technology pushes the offering forward. Also notice that needs direct the technology. There are multiple needs directing the development of the software. When desire enters into the picture it continues to push the offering forward. Need is no longer a factor because the basics of the offering have been established. As the technology grows in popularity and competition enters, continuing to define the offering, the only word that plays a pivotal role in the emergence of technology is ability. If the creator of the offering doesn't have the ability to develop the technology, the productivity level will fall and eventually fail to be a viable offering. This has been illustrated in Figure 18

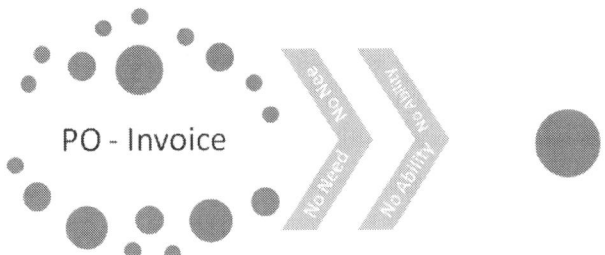

Figure 19

Need, ability, competition and desire will continue to be the catalysts of automated AP and the consumer will be the driver. Just as with other technology-driven industries. Given this perceptive, it is difficult to evaluate automated AP for the future; however I'll give to my best shot. Because of the recent stability of automated AP platforms, I believe most of the developments and improvements will be more focused on the process than the technology. Of course, there will be no shortage in technology improvements. Here are some improvements separated into technology and process.

Technology:

1. **Vendor** – In the future the vendor will have greater ability to provide your organization's automated AP system with multiple file formats and multiple distribution methods. A conversation with a new vendor about payment will not have anything to do with a bill to address, but more of a bill to type. The vendor would ask, "Do you want an, XML, CSV or a Text file?" or "Do you want the invoice sent to an

FTP site, email or transmitted through a web service call?" Software companies will emerge to create and manage technology that helps the vendor get up to speed (so to speak).

2. **Capture** – There are going to be a few changes on the capture side of technology. OCR (along with additional recognition software to rival OCR) will improve and become better at converting paper to data. However, the big change in this technology will be its decline. Vendors will become more adept at sending electronic information and AP systems at receiving the information thus cutting out paper all together.

3. **Workflow** – There is a lot of mystery around workflow. Within automated AP, workflow is the engine that drives the approval process. The term workflow is used to describe a process as well as a technology. In this case, in the future, workflow will become more conditionally responsive. Popular conditions with current technology are dollar amount or budget. In the future, multiple conditions for triggering workflows will be available. These conditions include codes, vendors, dates, budget, contract, purchase order, location, department, projects, entity, business unit, sub account, dollar amount, user, attention to, vendor discount, past due amounts... (just to name a few). It is important to point out that there are many solutions available in the market today, but not all of the above conditions have been developed. In the future, the evolution and power of workflow will not be the availability of the workflow triggers, but the use of workflow triggers in combination with each other. For example, a workflow step in which the approval

codes, using a certain code or set of codes, will trigger an additional approver with the exception of a certain vendor type or classification with an additional exception of a particular location or project. Currently this type of workflow technology does exist; however the change and the real improvement will come in how these conditions are administered and created. There will be a move for less development and more configurations within the accounting department and not IT.

4. **Reporting** – Reporting will be a conduit between technology and process. On the technology side the ability to create quicker and more accurate reports from multi-layer search tools. (Review reporting under process in this chapter for the good stuff)

5. **Accounting/ERP Systems** – For the most part (as of 2010) automated AP is a third party management product. You purchase the technology from a company that only does automated AP. In the future, accounting or ERP systems are going to add the ability to pull in vendor invoices, convert the invoices to data, have the invoices go through an approval process, then post them to GL. Be warned that when accounting systems start to do this the technology will not be as strong as the third party offerings due to the need for the accounting systems to hedge the overall expense of the accounting system. For the near future automated accounts payable is too complicated for an accounting system to do well. The tipping point for accounting or ERP systems will come when the automated AP is the norm and

the replacement wars begin (which is my next book... "Replacement Wars).

Process:

1. **Reporting** – Reporting will become less data and more business intelligence. Business intelligence in the using of data to make decisions. One of the basic principles I relate to companies that have paper processes and are looking to automate is "you don't know what you don't know". There is a lot of dysfunction in the paper process. Multiple checks and balances that are unnecessary in an automated world. Business intelligence will not only help uncover that paper dysfunction, it will help the company improve their AP process to a point where it will be unrecognizable to its previous process. Accounts payable data will also be combined with more accounting data and other data throughout the rest of the company, to make even larger, more critical processes and business decisions.

2. **Six Sigma** – Well it is about time...right! The greatest modern day process improvement miracle superhero is coming to an accounts payable department near you. Six Sigma was developed by Bill Smith[7] while working for Motorola. The point behind Six Sigma is to improve processes. The conception and application created works very well for manufacturing companies. However, not so well when accounts payable is paper. Paper creates too many variables. With the advent and mastery and widespread use of automated accounts payable, Six Sigma will apply very well, as will another

---

[7] http://www.isixsigma.com/index.php?option=com_k2&view=item&id=1505:the-history-of-six-sigma&Itemid=156

process design tool call Lean Process. In the future Lean Process and Six Sigma will play a pivotal role in accounts payable automation. Here are a few Six Sigma goals:

- Efforts to create practicable results for the business' success.
- Processes have characteristics that can be measured, analyzed, improved and controlled.
- Gaining quality is a total team effort from the top down and back up.

3. **Cash Management** – The cash management process is generally complete after the invoice has been approved and before the payable is promoted to the GL to be paid. In the future cash management will start when the PO is created or the invoice is transmitted from the supplier. Through the use of business intelligence the notion of cash management will be less checking a bank account to make sure funds are available and more of an algorithm of terms, discounts, available cash, approval cycles, internal rate of return and payment cycles. This will be a significant change for the better, as companies needs to manage change become more and more important.

There are more processes and technology improvements the future has to hold, I'm just providing some of the highlights. As I have written earlier in this chapter; need, ability, competition and desire will be the ultimate transformation of automated AP. These factors make the future hard to predict. Think in terms of our current lives (2010), twenty years ago - where was the Internet? Or even fifteen years ago - how important was a handheld wireless device to our businesses? Two years ago there were no IPads (what a tragedy). Not knowing the game changers that are under development or have yet to be thought of restricts our talking about future. It is safe to write that, in the future, paper is going away, AP is going to be automated and filing cabinets will be those things sold at yard sells with your children's children saying, "Dad. What is that?"

Conclusion

Hope you enjoyed the show!

My goals for writing this book are.

1. Write a book people will want to read.
2. Write a book people will read.
3. Write a book people will understand (non-technical).
4. Write a book people will use.

The first and second goals seem similar; however, inserting "want to" puts the emphasis on the subject matter: automated accounts payable. It is a very specific subject for a department within a department. I have kept Automated AP as the subject of focus for the benefit of the reader. I could have written a book about automation or using the internet to automate tasks, but the benefits would have been watered down. It is important to understand that, if the subject has not been published in the past, it is either ahead of its time or completely useless. I am open to either option. I have observed the pain that companies endure in deciding which solution to choose, what type of technology to leverage, build or buy and the projects that kick off and die a painful death. These signs point to the subject of automated AP being needed and useful. Dropping the "want to" from the goal my second desire is to have the book read. It is important to me that every paragraph of every chapter has meaning and keeps you reading to learn every step of the way. That is I chose my pitfall subjects carefully and didn't write long stories about my kids or repeated points over and over.

I want the reader to be the frontline decision maker, gaining insight and applying principles in their fight to improve accounts payable. I also want to write this book so that an accounting department employee (at any level) could be assisted and armed with good questions and knowledge about automation as they talk to internal IT and potential automation vendors.

Lastly, each chapter has at least one worksheet to help you personalize the material and use the results as a mapping system in going from paper to electronic data or improve your ingesting process. Too lofty? Not really. I desired to write this book from a perspective of use, not of theory. It is not a book that an MBA program will dive in and analyze, pulling out theoretic business principles. It is not a book that will sit on the desk of every CEO in the country as required reading for success (although CEOs should read it). It is a book for a person working diligently for an organization and desiring to make a greater impact in that organization. This is a book that is written out of experience and frustration - frustration in wanting to share this information and not having the time or vehicle to do so. I have perspective, lots of information (based on experience), and a desire for you (the reader) not to fall into the pit. In the pit are all the half-baked, poorly executed, left out in the rain, career killing projects that never had a chance. Staying out of the pit are those projects that live and breathe and produce high fives every time they are mentioned. These are the types of projects that a CEO and executive love to take credit for. Follow the course:

1. Current Environment – **Falling before you get started** - *Pitfall – understanding who your company is and where you are will keep you out of the pit.*
2. The Selection – **Falling for the wrong one** - *Pitfall – making assumptions with vendor selection will surely put you in the pit.*
3. The Organization – **Falling over yourself** - *Pitfall – doing your homework will keep you out of the pit.*

4. The Project – **Falling out of control** - *Pitfall – perception is not reality, keeping the project organized and in control will keep you out of the pit.*
5. The Vendors – **Falling for another** - *Pitfall – Poor or misdirected communication can throw you in the pit!*
6. The Training and Going Live – **Falling and can't get up** - *Pitfall – keeping training in the proper perspective will keep you out of the pit.*
7. The Evaluation – **Falling in Love** - *Pitfall – AP Automation is not an event. In order to stay out of the pit you need to be continually improving.*
8. The Future – **Falling over and over again** - *Pitfall – Having an idea of what is to come will keep you out of the pit.*

Nobody likes to trip and fall. There are many ways that people handle falling; jump up and pretend it never happened, stay where you fell – cry for help – and have everyone take pity on you, I can't move call 911, roll for a while (comedic or not), or try to take someone down with you... Whatever type of fall you do - let's face it, wouldn't you rather save yourself the embarrassment and not fall at all? If you knew the next step you were about to take would be the one in which your ankle gives way and you tumble down the stairs, wouldn't you tread lightly? Better yet, if someone was able to advise you which shoe to wear each day to make falling impossible would you take the advice?

This is the Tough Part!

The difficult part about all this wonderful information is, the problems being solved (getting paper out of the accounts payable process) are not that large in scale to the rest of the company. For example, I was talking with a client about centralizing their process. In its current state their process had multiple entities processing invoices. Through skillful analysis I was able to show him that he would have the ability to cut or reassign two people working in the process. He said it would be a difficult sell internally to the CFO. He went on to say that my challenge in expanding our relationship was going to be in the CFO's indifference to the problem. I saw the problem as having too many people in AP. When there are too many people, there are too many salaries. Salaries are the most expensive aspect of a business. My client explained that I would help the company save (net) $60,000 (I thought I had done a great thing). He went on to explain that the CFO of his company is not that interested in saving $60,000. The CFO's biggest concern was to refinance $68 Billon dollars' worth of mortgages in the most difficult lending market in years. I started to think about what he told me and it led me to hundreds of other conversations where indifference to what I was proposing was the biggest hurdle. The indifference sounds like this, "It's not so bad... the bills are getting paid. Sure, there are a few late fees and sometimes an invoice gets lost. There are a few evenings and weekends of overtime, but that's why I am paying accounting to get this work done. And besides, AP is a necessary evil. It's not like it creates revenue or generates sales." All of this is true. AP is a department within a department that is considered a pure expense, not an investment. So, what is my answer to this? AP Automation is the way of the future. That is true, in a future time there will be no paper invoicing, nothing to file and nothing to be entered into an AP system. There will also be no checks to sign, no paper to attach as back up and no one will be needed to do that type of work. Based on the last sentence here is my question to you; "Where do you want to be in the conversation of accounts payable automation?" Knowing that this is the wave of the future, do you want to be the AP department that has to change, or wants to change? Do you want to be the AP department with all the typical dally duties plus a yearlong automation cleanup project? I have outlined all the pitfalls in going from paper to automation for those people who want to proactively change and are not sure where to start or are worried about what they do not know. For those that are waiting until they have to change; really think about the spirit of change when you need to and when you do not need to. I love the employment recruiter's line, "You always make your best decisions when you don't have to make a decision." My advice is to automate sooner than later.

I Loved That Game!

I remember the day my family got an Atari 2600. We were a slow adaptors; my dad was waiting for it to drop under $100 before making the investment. That day I played Defender until the images were so imprinted into my mind, when I closed my eyes that night I could still see the game and its actions clearly. Months later, and through a lucrative trade with Kyle Keeter's sister, my brother and I acquired the game Pitfall (Fig. 9-1). It was a game about a man who ran through the jungle grabbing vines, swinging over pits with alligators (you had to jump on the backs of their heads too) to gain prizes (jewelry, gold...) If you missed the rope you were sent into the pit where with only a skillful, well-timed jump you could avoid getting stung by a scorpion. As I wrote this book I thought of that poor one dimensional fellow who I sent to his death many times because of poor execution and planning. Not so with automated accounts payable. This book will keep you out of the pit with the right shoes on, so you will not fall into the pit. Also, in reading this book you will get the prize at the end. It is not jewelry and gold but a product and processes that make an impact on your organization's bottom line... and you will be the hero!

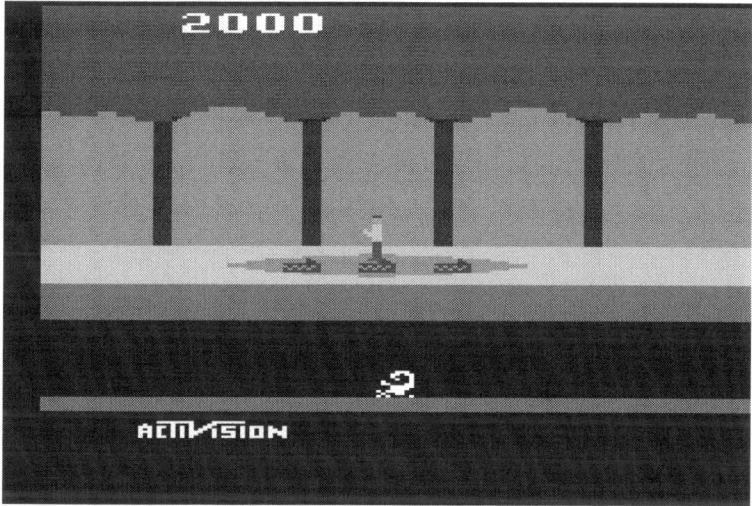

Figure 20

Acknowledgements

This is the most difficult step of the process. Have you ever been in front of a group of people and felt the overwhelming humble desire out of the goodness of your heart to thank the people that deserve thanking? Then, in the middle of naming names, you realize that there is no way you will be able to give credit where credit is due... This is followed by a sinking feeling that you will leave someone out that really deserves to be recognized... Well this is my attempt to make sure everyone gets the proper recognition that helped in writing this book.

Frist big thanks to my editors Rachel Stroud and Jan Skinner. It is a real talent to make a kid in high school that was in the half of the class that made the top half possible read like a pro. My wife too Adella for helping me re-write and for all of the wonderful support. I want to thank my children too, especially my oldest Kyle who told his sisters, Emma and Ada and his brother Ed that "Dad was writing a book and it is going to make him famous". I didn't have the heart to tell him that no one has ever gotten famous writing books on AP. Please don't tell him.

I also want to thank Mike Praeger and David Miller for allowing me the experiences and the room to express all of the ideas in this book through AvidXchange.

Thanks to Bill Whitley for giving me the outline framework and most importantly the motivation to do this book. Also, thanks Bill for the encouragement that a book like this was not only needed but useful.

Thanks Scot and Kris for all of the chili fests in 1994! (Scot put that in his book so I wanted to pay him back!)

Here is the fun part... trying to name the people who were part of all of the experiences that make the ideas in this book possible (here goes... hold on)

Chris Freddy Rich Alan Kristen Christa Greg Patrick Brian Brian Joe Bob Rich Tom Craig Tom 87 Jodi Heather (especially Heather) Jay Tim Jodi Ashley Josh Charles Ron Daryl Robert Britt Hunter Alex Amy Angela Anil Betty Bill (not the Gorilla) Dameon Donald Greg Irene Jamie Jason Jeff Joe Joel Karen Manoj Mark Megan Nora Paul Phillip Becky Sudhir Tim Bob Ben David Gary Brett Brian Pat Terry Richard Mohan Tina Adam Lauren Diana Jeff Doug David Dave Robert Richard Deborah Mike Melanie Nicole Joe Chris Matthew Katy Kim Keith Rita Stuart Peter Kristin David Jon Michael Bill Rhonda James Adam Brandon Brooke Melanie Randy Alan Becky Bjorn Bryce Chris Dan Jim Shannon Ted Patrick Richard Wilson Patrick Elaine Scott Millie Beth Irfan Chad Karla Scott Michael Scott Douglas Steve William Bill Jim Alicia Justin Steve Andy Frederick Kenneth Michael Brent Dan Steven Donna Scott Nathan Greg Don David Tonya Jon Robin Edward Greg Dean Cynthia Erin Annmarie Art Deanna Barb William Greta Bobby Greg Patrick Kevin Frank Elizabeth John Daniel Susan Michael Ross Gary Oliver Rory Michael Charles Michael James Barry Joel Douglas Nicholas Timothy Jeff Dana Ralph Jennifer Abby Carmen Harold Matthew Mary Tony Kathryn Kevin Michael William Charmaine Lilian John Karen Ray Evrett Rosemary. Brian Bruce Fran Barry Aaron John Paul Andy David Doug Andy

8 PITFALLS OF ACCOUNTS PAYABLE AUTOMATION

First Edition 2011

David Nicole Paul Michael Richard Richard Gary Charles Craig Patrice Robert Hugh Marc Gregory Andrew Jon Jon Sheryl Michelle Joe Pam Ed Steve Warren Richard Suzanne Steve Suzanne Richard Chris Roger Jeff Cathy Tanya Veronica Penny Steve Shelby Jerome John Tory Patrick Steven John Duncan Rebecca Miriam Edward Lee David Bruce Paul Riki Henry Neil Tyler Mark Bill Charles Burton Ron Christopher James Scott Bill Leigh Arthur Chon Bill Eva Terrie Thomas Shawn Charles Wesley John Ray Doug Clayton Dorothy Mark Kent Louis Thomas David Scott Eric Jack Stephen Jay Anne J D Mark Hampton Tracey Kevin Chris Philip Gerald Joseph David Oliver Scott David Dan Dave Steven Timothy Philip Steve Brian Robert Stephen Tamara Robert Ben Maury Lance Jack Geoff Elizabeth Jane Paul Alejandra Steve Dave Jessica Kim Julie Andy Judy Amy Philip Kim Julie Al Chuck Daniel David Lisa Moe Benjamin Tim Jack Elizabeth Cindy Greg Kim Rick Dana Connie Mark Kim Dana Toni Mary Sandra Terry Jerry Bill Chad Stephen Macy Edward Dianne Susan Robert Jeff Misty Francisco Pamela Glinda Sylvia Jasper Angeliegh Jon Karen Christine Jim Ted Kelly Monroe Ed Diane Melinda\ Christie David Jerry Barbara James Jeff Gary Aaron Christopher John Christine Mark Bob Tracy Jennifer Timothy Nick Amanda Amy Elizabeth Mary Chris Kevin Janet Brett Anne Bill Maggie Dionne Helena Anne Aaron David Donald Terry Matt Sean John Bobbie Marci Vivian Priscilla Chris Frank Richard Ken Vicki Nicholas Carrie Grace Arthur William Steven Kevin Joseph Joy Amy Tracey Renata Alan Dawn Keith John Kent Cindy Debbie Beth Maureen Joanna Jessica Vicki William Richard Donna Debbie Rebecca Kevin Linda Sean Jackie Amy Katherine Mark Alex Wendy Roy Michael Kristy James Jose Theresa Michael Karen John Paul Dallas Linda Debi Anne Robert Paul Matt Todd John William Thomas William

8 PITFALLS OF ACCOUNTS PAYABLE AUTOMATION

First Edition 2011

Charles John Barbara Craig Ray Kelly Matt Jennifer Tom Tom Rachel Sandra Ed John Scott Cindy Derik Kay Charles Chip Tim Kenneth Sandy Ben Jason Todd Shelley Clay Tina Kathy Suzanne Jeff Tom Elizabeth David Patrick Janet Mary Mike Lisa Lisa Mark Dave Lynn Louie Kdy Thomas David Herbert Rich Frederick Scott Russ Catherine Gary Dennis Walter Tracy Janet Shelley Carol Joel Graham Teresa Michal Kenneth Bill Michael Robert Chris Norman Anthony Mark or and Mom and Dad!

Can you believe all of those names? I know I left someone out. That's why I did first names. There are a lot of people that helped me with this book. To be clear most of the people in the list are not aware that they helped, but in order to write a book about real life experiences, you need to have a lot of real life experiences. Each person in the list I have either worked with to automate their Accounts Payable processes or we worked together as a service provider to offer automated technology.

Finally, I prayed a lot about this book and I asked God to guide me in producing the most wisdom I could possibly produce on the subject... I hope I have done Him proud and the ultimate acknowledgement has to go to Him and His kingdom!

See Ya paper!

Chris

Glossary

Term	Definition
AP Automation	The independent moving of liabilities through approvals to payment.
	There are three main components to make AP Automation happen.
	1. 100% Electronic Invoices
	2. Event Driven Workflow
	3. Reporting Layer to Track all Actions
Best Practice	A teaching tool that over time, effort and cost can be done in an optimized state.
Document Management	An electronically storing a file (such as an invoice). Automation is the creation of an electronic process (that usually replaces a manually process).
Software as a Service (SaaS)	A single instance of the software to manage multiple user base that is hosted by the service provider
Application Service Provider (ASP)	A single database and instance of the software of a single user base that is hosted by the service provide

Self-hosted	A single database and instance of the software of a single user base that is hosted by the client or internal user's organization
ROI	Return on Investment. This is the cost of the new process or technology minus the savings of the new process or technology
Sarbanes Oxley (SOX)	The **Sarbanes–Oxley Act of 2002** (Pub.L. 107-204, 116 Stat. 745, enacted July 30, 2002), also known as the 'Public Company Accounting Reform and Investor Protection Act' (in the Senate) and 'Corporate and Auditing Accountability and Responsibility Act' (in the House) and commonly called **Sarbanes–Oxley**, **Sarbox** or **SOX**, is a United States federal law enacted on July 30, 2002, which set new or enhanced standards for all U.S. public company boards, management and public accounting firms. It is named after sponsors U.S. Senator Paul Sarbanes (D-MD) and U.S. Representative Michael G. Oxley (R-OH).
Re-Work	Having to repeat the same work more than once.
Workflow	Predetermined electronic approval process.

Roles	Similar to job titles, are groups of people that have a certain set of permissions and abilities.
Approver	A person that is doing the work within the workflow process.
EIPP	Electronic Invoice Presentment and Payment
CSV	Comma Separated Values
FTP	File Transfer Protocol
TIFF	Tagged Image File Format
PDF	Portable Document Formant
XML	Extensible Markup Language
OCR	Optical Character Recognition
EDI	Electronic Data Interchange
EPD	Early Payment Discounts

Made in the USA
Charleston, SC
12 June 2013